Managing Food Safety Risks in the Agri-Food Industries

Managing Food Safety Risks in the Agri-Food Industries

Jan Mei Soon • Richard Baines

CRC Press
Taylor & Francis Group
Boca Raton London New York

CRC Press is an imprint of the
Taylor & Francis Group, an **informa** business

CRC Press
Taylor & Francis Group
6000 Broken Sound Parkway NW, Suite 300
Boca Raton, FL 33487-2742

First issued in paperback 2017

© 2014 by Taylor & Francis Group, LLC
CRC Press is an imprint of Taylor & Francis Group, an Informa business

No claim to original U.S. Government works

ISBN-13: 978-1-4665-0950-4 (hbk)
ISBN-13: 978-1-138-19992-7 (pbk)

Visit the Taylor & Francis Web site at
http://www.taylorandfrancis.com

and the CRC Press Web site at
http://www.crcpress.com

Contents

Preface

Modern farming practices are becoming increasingly intensive and consolidated in order to meet the needs of the rest of the supply chain. These stages involve more players (stakeholders) in the supply chain, not to mention issues of storage, transportation, and distribution prior to consumer receipt. This can be the trigger where microorganisms contaminate a large number of crops, livestock, or aquatic organisms. During the preparation of this proposal, Germany faced one of the largest and worst Shiga toxin-producing *Escherichia coli* O104:H4 outbreaks in history. After a long and arduous investigation, the source of contamination was identified as fenugreek sprouts. This example is used to emphasize how food production chains are becoming more complex due to many drivers including increasing and more intensive production, global sourcing, vertical integration along discrete chains, and consolidation of businesses at key stages of supply.

Therefore, the aim of this book is to inform readers of existing and emerging risks in the primary agri-food industries and ways to manage, reduce, or prevent the likelihood of risks from occurring.

This book covers three main primary production sectors—crops, terrestrial livestock, and aquaculture products—along with a chapter on game and wild fish catch. Under each sector, the readers are presented with existing and emerging food safety risks, challenges, and intervention strategies. A review of some of the significant (past) food safety hazards informs readers on how the agri-food industries had dealt with the problems, while discussions on present and emerging hazards will give the readers an overview of the current situation. Risk management strategies such as Good Agricultural/Aquaculture Practice, HACCP-based principles, and traceability are some of the main recommendations covered from Chapters 3–7.

Jan Mei Soon
Richard N. Baines

1

Introduction

During the initial conceptualization of this book, Germany faced one of the largest and worst Shiga toxin-producing *Escherichia coli* O104:H4 outbreaks in history. After a long and arduous investigation, the source of contamination was identified as fenugreek sprouts. What have we learned from this massive outbreak? How can we prevent or reduce foodborne outbreaks from recurring? What were the risks faced by consumers? Risk—a simple word, but relatively difficult to define. Risk is the probability of an adverse health effect and the severity of that effect due to a hazard in food. In other words, risk is the likelihood and seriousness of a hazard.

This book covers three primary production sectors—crops, terrestrial livestock, and aquaculture products, along with a chapter on wild game and fishery. Each chapter focuses mainly on microbiological and chemical (natural or man-made contaminants) hazards that occur at the farm level with the potential to cross-contaminate (e.g., pathogens in ready-to-eat fresh produce, *Salmonella* spp. in poultry) or bioaccumulate (e.g., polychlorinated biphenyls in farmed fish). Physical hazards on the other hand are not discussed as extensively as microbiological or chemical contaminants. This is due to the fact that most physical hazards in primary production crops such as fresh produce, meat, or fish products are eliminated or reduced through visual observation and are at times associated as part of the natural environment of the product.

Chapter 2 emphasizes the different types of risk assessment—qualitative, semi-quantitative, and quantitative risk assessments. Various risk assessment tools that have been developed and utilized in the food chain are also discussed. Chapter 3 focuses on horticulture crops. Microbiological contaminants and an outbreak investigation example of *E. coli* O157:H7 in bagged spinach is used to identify weak-point sources of contamination and to apply intervention strategies to reduce or prevent the same mistakes from recurring.

Chapter 4 discussed the main food safety hazards associated with beef, sources and transmission routes of pathogenic bacteria, and prevention strategies. This is followed by milkborne outbreaks under the dairy section. Farm risk factors associated with milk safety and intervention strategies to reduce contamination are also discussed.

In the following meat chapter, chicken is one of the most widely accepted muscle foods in the world. It is not a coincidence that poultry are considered a major source of human infection due to its huge consumption volume. The main microbiological risks associated with the broiler industry are

Campylobacter and *Salmonella*. In the next section, we shall focus on the prevalence of *Salmonella* in pig meat, on-farm risk factors, and intervention strategies.

Chapter 6 emphasizes the food safety risks, especially microbiological and environmental contaminants faced by captured fish, and potential parasites and diseases from the game and exotic meat industry. In addition to wild fishery, aquaculture production such as fish and shellfish farming are also associated with pathogenic bacteria and chemical contaminants. Intervention strategies including a HACCP-based plan for salmon production at marine sites is also shown in Chapter 7.

In Chapter 8, we shall explore the consumers' and agri-food stakeholders' risk perceptions and some of the best media for risk communication. The questions of how do public and experts perceive risks and how do we communicate effectively about food safety risks to consumers are also explored.

2

Risk Assessment

This chapter discusses the various risk assessment tools developed and utilized in the food chain. Different types of risk assessment—qualitative, semi-quantitative, and quantitative—advantages and disadvantages will also be evaluated.

2.1 Introduction

2.1.1 Hazards and Risk Assessment

This is an issue for the specialists in the fields of food spoilage, microbiology, toxicology, and alike. However, the majority of those involved in the industry are not specialists but are responsible for managing the safety of food, in this case beef. Therefore, specialist knowledge of hazards and understanding of the associated risks should be built into a framework for managing hazards and risks. Hazards may be divided into three main classes based on the interaction with the food product (Table 2.1). The presence of each may influence the type of assessments made to determine whether critical limits are exceeded, and the actions to be taken. In addition, the type of assessment may affect when food products that exceed critical limits are removed from the supply chain. Detecting the presence of both physical and chemical hazards can be built into continuous sampling at critical points in the process with contaminated materials removed immediately. In contrast, biological hazards often require a period of culturing and bioassay before critical levels can be determined; as a result, products may have moved further along the supply chain, and thus requires product recall.

2.1.2 What Is Risk?

Risk is "a function of the probability of an adverse health effect and the severity of that effect, consequential to a hazard(s) in food" (CAC, 2001). (See Figure 2.1.) Risk analysis is used to develop an estimate of risks to human health and safety, to identify and implement appropriate measures to control the risks, and to communicate to stakeholders about the risks and measures

TABLE 2.1

Hazards That Can Contaminate Foods, Types of Assessment, and Possible Actions to Be Taken

Nature of Hazard	Impact on Food Product	Assessment and Actions
Physical	Physical contaminant harmful, e.g., glass Mixed with, or covering, food product but no chemical reaction	Visual assessment Chemical assessment of contaminant Remove source of contamination Consider separation of contaminant from food product
Chemical	Chemical contaminant harmful to humans, e.g., animal drugs, pesticides Chemical interacts with food product to make it unsafe	Chemical assessment of contaminant or derivatives Remove source of contamination Reject and dispose of food product
Biological	Biological agent or its toxins harmful to humans, e.g., pathogens Colonization of food product causing spoilage or secondary contamination	Bioassay or culture techniques Remove source of contamination Reject and dispose of food product

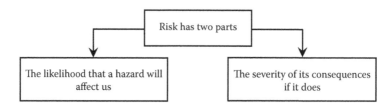

FIGURE 2.1
Risk is the likelihood and severity of a hazard. (From Sumner, J. et al. 2004. Application of Risk Assessment in the Fish Industry. FAO Fisheries Technical Paper 442, pp. 1–78. ftp://ftp.fao.org/docrep/fao/007/y4722e/y4722e00.pdf [accessed April 15, 2010].)

applied. Risk analysis is made up of three distinct but closely connected components—risk assessment, risk management, and risk communication—which have the overall objective to ensure public health protection (FAO/WHO, 2006; Toyofuku, 2006; Figure 2.2).

Risk assessment is the science-based component of risk analysis. Risk management is the component in which scientific information from risk assessment, economic, and social considerations are weighed to choose appropriate prevention and control points. Risk communication is the exchange of information and opinions throughout the risk analysis process among risk assessors, risk managers, consumers, industries, academia, and other interested parties (FAO/WHO, 2006). This chapter will focus on risk assessment with a special emphasis on the types of assessment and decision support tools used in this key area.

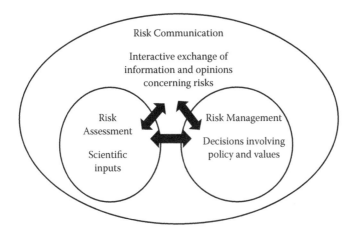

FIGURE 2.2
Components of risk analysis. (From FAO/WHO. 2006. Food Safety Risk Analysis. A Guide for National Food Safety Authorities. pp. 1–78. http://www.who.int/foodsafety/publications/micro/riskanalysis06.pdf [accessed April 5, 2010].])

2.1.3 Risk Assessment

Codex Alimentarius has defined risk assessment as a four-step scientifically based process of risk analysis consisting of hazard identification, hazard characterization, exposure assessment, and risk characterization (*Codex Alimentarius*, 1999). This definition recognizes that the ultimate goal of risk assessment process is to estimate the probability and severity of risk occurrence using qualitative and/or quantitative information (Coleman and Marks, 1999; Davidson et al., 2006) and subsequently to identify opportunities for intervention (Schlundt, 2000). It also stipulates that uncertainty is recognized and included in any estimates of risk (Davidson et al., 2006). Risk assessment in a food safety context has two meanings as suggested by Schlundt (2000). The generic meaning of risk refers to the measurement of risk and the identification of factors that influence it (Voysey and Brown, 2000). The specific (or formalized) meaning of risk assessment is the scientific evaluation of known or potential adverse health effects resulting from human exposure to foodborne hazards (FAO/WHO, 1995). Risk assessment and HACCP are related, but are fundamentally different processes. Similarities exist between the inputs in the first element of HACCP (hazard analysis) and risk assessment (hazard identification) (Coleman and Marks, 1999). Risk assessment provides support to risk managers in decision making and is divided into qualitative, semi-quantitative, and quantitative assessment (Sumner et al., 2004).

2.1.3.1 Qualitative Risk Assessment

According to CAC (2001), qualitative risk assessment is "based on data which, while forming an inadequate basis for numerical risk estimations, nonetheless, when conditioned by prior expert knowledge and identification of attendant uncertainties permits risk ranking or separation into descriptive categories of risk." Hence, qualitative risk assessment can assist a risk manager in priority setting and policy decision making (Coleman and Marks, 1999).

2.1.3.2 Semi-Quantitative Risk Assessment

Semi-quantitative risk assessment forms the bridge between qualitative and fully quantitative methods. Values can be represented with linguistic and/or numeric scales and some quantitative measures of risk are produced (Davidson et al., 2006).

2.1.3.3 Quantitative Risk Assessment

According to Voysey and Brown (2000), a Quantitative Risk Assessment (QRA) should be carried out wherever or whenever possible. However, if no data are available to make inferences from, then a quantitative risk assessment is not possible (Coleman and Marks, 1999). QRA are usually carried out to evaluate microbiological hazards. A quantitative microbial risk assessment (MRA) produces a mathematical statement that links the probability of exposure to an agent and the probability that the exposure will affect the test individual (Voysey and Brown, 2000). If a qualitative risk assessment has been done, the risk estimate will be a simple statement that the risk is high/medium/low. If it is a quantitative risk assessment, the risk estimate will be a number, such as predicted illnesses per annum in the population, or the probability of becoming ill from eating a serving of the food (Sumner et al., 2004). Several quantitative risk assessments for specific microbiological hazards in products such as *Escherichia coli* O157:H7 in ground beef hamburgers (Cassin et al., 1998), *Vibrio parahaemolyticus* in raw oysters (FDA, 2005), and *Salmonella* in whole chickens (Oscar, 2004) are shown (Table 2.2). Cassin et al. (1998) described the behavior of pathogens from production through processing, handling, and consumption to predict human exposure to *E. coli* O157:H7 from beef hamburgers. The exposure estimate was then used as input to a dose-response model to estimate the health risk associated with consumption. This model predicted a probability of hemolytic uremic syndrome (kidney failure) of 3.7×6.7^{-6} and a mortality of 1.9×10^{-7} per meal for the very young. Meanwhile, the FDA (2005) predicted the risk per serving of raw oysters as approximately 1×10^{-6} (equivalent to one illness of gastroenteritis in every 1,000,000 servings). Oscar (2004) also predicted 0.44 cases of salmonellosis per 100,000 consumers per chicken serving.

Quantitative risk assessments have the additional advantage of being able to model the effects of different interventions, and this is probably the greatest strength of QRA (FAO/WHO, 2006). However, the disadvantage of quantitative microbiological risk assessment is that it is time-consuming and

TABLE 2.2

Quantitative Risk Assessment Models for Food Pathogens

Pathogen	Food Commodity	Probability of Infection	Reference
Escherichia coli O157:H7	Ground beef hamburgers	3.7×6.7^{-6}	Cassin et al., 1998
Listeria monocytogenes	Soft cheese	1.9×10^{-9} to 6.4×10^{-8}	Bemrah et al., 1998
Bacillus cereus	Chinese-style rice	2.13×10^{-3} (if food is held at room temperature of 20°C)	McElroy et al., 1999
Listeria monocytogenes	Smoked salmon	1.3×10^{-4} (assuming only high-risk population)	Lindqvist and Westöö, 2000
Escherichia coli O157:H7	Raw fermented sausages	0.15% probability of detecting *E. coli* O157:H7 in 25 g sample	Hoornstra and Notermans, 2001
Vibrio parahaemolyticus	Raw oysters	1×10^{-6}	FDA, 2005
Salmonella spp.	Whole chicken	0.44×10^{-5}	Oscar, 2004

expensive. In addition, by the time QRA has been carried out, many food products may have left the food businesses and consumed.

2.1.4 Risk-Ranking Tools

Risk assessments vary greatly in their level of detail (McNab, 2003). Assessment may range from (1) qualitative opinions, rating risks as low, medium, or high (Soon et al., 2013); (2) well-documented, systematic qualitative assessment of the likelihood and consequence of defined risks (e.g., Huss et al., 2000), (3) semi-quantitative scoring and ranking systems (e.g., Ross and Sumner, 2002), and (4) complex, detailed, quantitative mathematical models that include Monte Carlo simulation and quantification of uncertainty (e.g., Cassin et al., 1998).

Food decision makers may require tools that enable them to: (1) identify the most significant risks from a public health perspective; (2) reduce risks, by taking into account the feasibility, effectiveness, and cost of possible interventions, and (3) allocate efforts and resources accordingly (Food Safety Research Consortium, 2003; Taylor et al., 2003). There are a number of decision support tools to assist in determining potential food safety risks and infectious diseases and 12 of these tools are summarized (Table 2.3).

The choice among qualitative, semi-quantitative, and quantitative approaches is largely driven by the objectives of the decision makers (i.e., a qualitative approach may be sufficient for their needs), and the availability of data and expertise of the analysts (Peeler et al., 2007). The authors determined that the additional benefit of a quantitative compared to a qualitative analysis should also be judged by whether the basis for decision making is

TABLE 2.3

Risk-Ranking Tools

Risk-Ranking Tools	Type of Assessment	Description	Country of Origin	Reference
Simple scoring scheme	Qualitative	Divided into six categories of risk factors and characteristics of known hazards in seafood. Based on a survey of outbreaks of illnesses attributed to different seafood, a product was scored as either + or − for each category and then ranked according to the total number of +'s.	—	Huss et al., 2000
The Business Food Safety Classification Tool	Qualitative	Provides guidance on the allocation of Australian food business sectors into categories based on their likelihood of contributing foodborne disease and the potential magnitude of that contribution.	Australia	Department of Health and Ageing, 2007
Risk Categorization Model (RCM) for Food Retail/ Food Service Establishments	Qualitative	Categorizes food establishments so that authorities can give greater attention to those with the greatest potential risks to consumers.	Canada	Canadian Federal/ Provincial/Territorial Committee on Food Safety Policy (FPTCFSP), 2007
Hygiene Risk Assessment Model (HYGRAM) system	Semi-quantitative	Hygiene Risk Assessment Model (HYGRAM) was developed for small- and medium-sized enterprises. HYGRAM consists of background information to allow users to key in their company's details, hygiene and hazard databank to generate results.	Finland	Tuominen et al., 2003
Foodborne Illness Risk Ranking Model (FIRRM)	Semi-quantitative	Allow users to rank pathogen/food risks using five criteria: number of cases, hospitalizations, deaths, economic costs of health outcomes and loss of Quality Adjusted Life Years (QALY).	—	Batz et al., 2005; FAO/ WHO, 2006; Food Safety Research Consortium, 2003

Continued

| Risk Ranger | Quantitative | A food safety risk calculation tool to determine relative risks from different product, pathogen and processing combinations. A risk estimate is calculated, scaled between 0 and 100, where 0 represents no risk and 100 represents that the meal contains a lethal dose of hazard. | Australia | Food Safety Centre, 2010; Ross and Sumner, 2002 |
| CARVER + Shock Tool | Qualitative | A food defensive tool adapted from the military for use in the food industry. The tool can be used to assess how vulnerable a system or infrastructure is to an attack and focus resources on protecting the system. CARVER is an acronym for the following six attributes used to evaluate the attractiveness of a target for attack:

• Criticality—measure of public health and economic impacts of an attack
• Accessibility—ability to physically access and egress from target
• Recuperability—ability of system to recover from attack
• Vulnerability—ease of accomplishing attack
• Effect—amount of direct loss due to attack
• Recognizability—ease of identifying target
• Shock—Health, economic, and psychological impacts of an attack within the food industry | United States | FDA, 2010 |

TABLE 2.3 (*Continued*)

Risk-Ranking Tools

Risk-Ranking Tools	Type of Assessment	Description	Country of Origin	Reference
Stepwise and Interactive Evaluation of Food Safety by an Expert System (SIEFE)	Quantitative	Developed for quantitative risk assessment associated with microbial hazards for food products and production processes. SIEFE is interactive and is best used by experienced microbiologists, as they are able to make the best use of their knowledge and interpret the system's estimates critically.	—	van Gerwen et al., 2000
Food Safety Universe Database (FSUDB)	Semi-quantitative	To assess and rank food safety risks across various foods and hazards, at various points along the food chain. It assesses risk based on consumption patterns and estimates risks per serving.	Canada	McNab, 2003
Swift Quantitative Microbiological Risk Assessment (sQMRA)	Quantitative	A simplified QMRA model aimed at comparing the risk of pathogen-food product combinations using Microsoft Excel.	—	Evers and Chardon, 2010
Import Risk Analysis (IRA)	Semi-quantitative	Used mainly in the assessment of diseases in the field of animal health. IRA is used to assess the relative importance of different routes of introduction and spread of diseases in farms.	—	Covello and Merkhofer, 1993; Peeler et al., 2007
Safe produce assessment	Qualitative	Assess farm food safety risks based on specific topics such as manure, irrigation water, and workers' hygienic practices.	United Kingdom	www.safeproduce.eu

significantly improved. If there is no data available, then a quantitative risk assessment would not be possible. Qualitative risk assessment can assist a risk manager, to a certain level, in priority setting and allocation of resources (Coleman and Marks, 1999). The selection of different levels of risk assessments is shown (Figure 2.3). The additional benefit of a quantitative compared to a qualitative analysis should also be judged by whether the basis for decision making is significantly improved (Peeler et al., 2007).

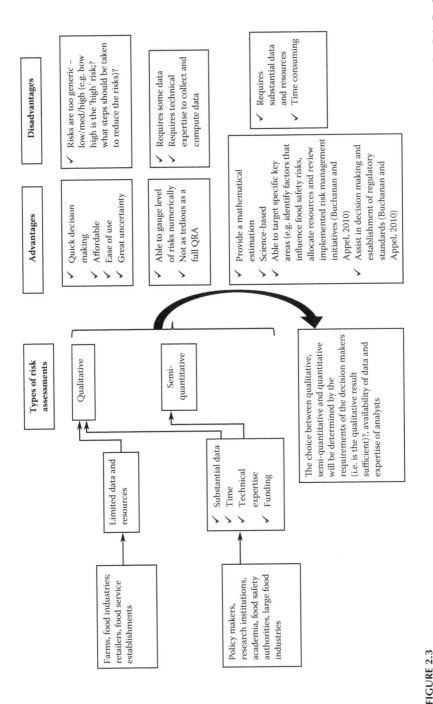

FIGURE 2.3
Selection of different levels of risk assessments. (From Manning, L. J. and J. M. Soon. 2013. Mechanisms for assessing food safety risks. *Brit. Food J.* 115(2).)

References

Batz, M. B., M. P. Doyle, J. G. Morris et al. 2005. Attributing illness to food. *Emerg. Infect. Dis.* 11: 993–999.

Bemrah, N., M. Sanaa, M. H. Cassin, M. W. Griffiths, and O. Cerf. 1998. Quantitative risk assessment of human listeriosis from consumption of soft cheese made from raw milk. *Prev. Vet. Med.* 37: 129–145.

Buchanan, R. L. and B. Appel. 2010. Combining analysis tools and mathematical modeling to enhance and harmonize food safety and food defense regulatory requirements. *Int. J. Food Microbiol.* 139: S48–S56.

CAC. 2001. Principles and guidelines for the conduct of microbiological risk assessment (CAC/GL 30-1999). *Food Hygiene Basic Texts*, Second Edition. ftp://ftp.fao.org/docrep/fao/005/Y1579e/Y1579e.pdf (accessed April 9, 2010).

Cassin, M. H., A. M. Lammerding, E. C. D. Todd, W. Ross, and R. S. McColl. 1998. Quantitative risk assessment for *Escherichia coli* O157:H7 in ground beef hamburgers. *Int. J. Food Microbiol.* 41: 21–44.

Codex Alimentarius. 1999. Principles and Guidelines for the Conduct of Microbiological Risk Assessment. CAC/GL-30 (1999) [online] Available from: www.codexalimentarius.net/download/standards/357/CXG_030e.pdf (accessed March 30, 2010).

Coleman, M. E. and H. M. Marks. 1999. Qualitative and quantitative risk assessment. *Food Control* 10: 289–297.

Covello, V. T. and M. W. Merkhofer. 1993. *Risk Assessment Methods: Approaches for Assessing Health and Environmental Risks.* New York: Plenum Publishing, New York.

Davidson, V. J., J. Ryks, and A. Fazil. 2006. Fuzzy risk assessment tool for microbial hazards in food systems. *Fuzzy Set Syst.* 157: 1201–1210.

Department of Health and Ageing. 2007. Business Sector Food Safety Risk Priority Classification Framework. Australian Government, Department of Health and Ageing.

Evers, E. G. and J. E. Chardon. 2010. A swift quantitative microbiological risk assessment (sQMRA) tool. *Food Control* 21: 319–330.

FAO/WHO. 1995. Application of Risk Analysis to Food Standards Issues. Report of the Joint FAO/WHO expert consultation, Geneva, Switzerland, March 13–17, 1995. http://www.who.int/foodsafety/publications/micro/en/march1995.pdf (accessed February 24, 2010).

FAO/WHO. 2006. Food Safety Risk Analysis. A Guide for National Food Safety Authorities. pp. 1–78. http://www.who.int/food safety/publications/micro/riskanal ysis06.pdf (accessed April 5, 2010).

FDA. 2005. Quantitative Risk Assessment on the Public Health Impact of Pathogenic *Vibrio parahaemolyticus* in Raw Oysters. Center for Food Safety and Applied Nutrition, Food and Drug Administration.http://www.fda.gov/Food/Science Research/ResearchAreas/RiskAssessmentSafetyAssessment/ucm050421.htm (accessed April 8, 2010).

FDA. 2010. CARVER Software. http://www.fda.gov/Food/FoodDefense/CARVER/default.htm (accessed January 3, 2011).

Food Safety Centre. 2010. Australia's Food Safety Information Portal. Risk Ranger. http://www.foodsa fetycentre.com.au/riskranger.php (accessed April 5, 2010).

Food Safety Research Consortium. 2003. Purpose and Key Attributes of the Foodborne Illness Risk Ranking Model. http://www.thefsrc.org/Model%20Purpose%20 and%20Key%20Attributes%20FORMATTED.pdf (accessed April 5, 2010).

FPTCFSP. 2007. Risk Categorization Model for Food Retail/Food Service Establishments. Health Canada, Federal/Provincial/Territorial Committee on Food Safety Policy. http://www.hc-sc.gc.ca/ahc-asc/alt_formats/hpfb-dgpsa/ pdf/pubs/risk_categorization-categorisation_risques-revised_revisee-eng.pdf (accessed April 5, 2010).

Hoornstra, E. and S. Notermans. 2001. Quantitative microbiological risk assessment. *Int. J. Food Microbiol.* 66: 21–29.

Huss, H. H., A. Reilly, and P. K. B. Embarek. 2000. Prevention and control of hazards in seafood. *Food Control* 11: 149–156.

Lindqvist, R. and A. Westöö. 2000. Quantitative risk assessment for *Listeria monocytogenes* in smoked or graved salmon and rainbow trout in Sweden. *Int. J. Food Microbiol.* 58: 181–196.

Manning, L. J. and J. M. Soon. 2013. Mechanisms for assessing food safety risks. *Brit. Food J.* 115(3)460–484.

McElroy, D. M., L.-A. Jaykus, and P. M. Foedgeding. 1999. A quantitative risk assessment for *Bacillus cereus* emetic disease associated with the consumption of Chinese-style rice. *J. Food Safety* 19: 209–229.

McNab, B. 2003. Food Safety Universe Database. A Semi-Quantitative Risk Assessment Tool. Ontario Ministry of Agriculture and Food https://ozone.scholarsportal. info/bitstream/1873/6231/1/10318750.pdf (accessed April 9, 2010).

Oscar, T. P. 2004. A quantitative risk assessment model for *Salmonella* and whole chickens. *Int. J. Food Microbiol.* 93: 231–247.

Peeler, E. J., A. G. Murray, and A. Thebault. 2007. The application of risk analysis in aquatic animal health management. *Prev. Vet. Med.* 81: 3–20.

Ross, T. and J. Sumner. 2002. A simple, spreadsheet-based, food safety risk assessment tool. *Int. J. Food Microbiol.* 77: 39–53.

Schlundt, J. 2000. Comparison of microbiological risk assessment studies published. *Int. J. Food Microbiol.* 58: 197–202.

Soon, J. M., W. P. Davies, S. A. Chadd, and R. N. Baines. 2013. Field application of farm food safety risk assessment (FRAMp) tool for small- and medium-fresh produce farms. *Food Chem.* 136: 1603–1609.

Sumner, J., T. Ross, and L. Ababouch. 2004. Application of Risk Assessment in the Fish Industry. FAO Fisheries Technical Paper 442, pp. 1–78. ftp://ftp.fao.org/ docrep/fao/007/y4722e/y4722e00.pdf (accessed April 15, 2010).

Taylor, M. R., M. O'K. Glavin, Jr., J. G. Morris, and C. E. Woteki. 2003. Food Safety Updated: Developing Tools for a More Science- and Risk-Based Approach. http://www.milbank.org/reports/2003foodsafety/030731foodsafety.html (accessed April 5, 2010).

Toyofuku, H. 2006. Harmonization of international risk assessment protocol. *Mar. Pollut. Bull.* 53: 579–590.

Tuominen, P., S. Hielm, K. Aarnisalo, L. Raaska, and R. Maijala. 2003. Trapping the food safety performance of a small or medium-sized food company using a risk-based model: The HYGRAM® system. *Food Control* 14: 573–578.

van Gerwen, S. J. C., M. C. te Giffel, K. van 't Riet, R. R. Beumer, and M. H. Zwietering 2000. Stepwise quantitative risk assessment as a tool for characterization of microbiological food safety. *J. Appl. Microbiol.* 88: 938–951.

Voysey, P. A. and M. Brown. 2000. Microbiological risk assessment: A new approach to food safety control. *Int. J. Food Microbiol.* 58(3): 173–179.

3

Managing Risks in the Horticultural Sector

3.1 Introduction

As more foods are purchased away from home or purchased in prepared forms, consumers can exercise less control over key food safety control points (Isshiki et al., 2009). This trend has led to a demand for a wider range of products with less preparation and cooking at home. Indeed, such trends are reflected in the increase of salad bars, sandwiches, juices, and minimally processed foods that are ready to eat. In this context, we define minimally processed foods as foods that are produce commodities that are eaten raw and have not received a formal process or treatment to reduce pathogenic bacteria, their spores, or toxins to a safe level. Matthews (2006) argued that the number of food poisoning cases or outbreaks due to produce contamination has increased, since fruits and vegetables are mostly produced in a natural environment and are vulnerable to contamination. Therefore, the consumption of ready-to-eat produce or minimally processed fresh-cut produce, if suitable safeguards are not in place, represents a new challenge to food safety (Bhagwat, 2006).

The consumption of fresh-cut or minimally processed fruit and vegetables has undergone a sharp increase (Abadias et al., 2008). The World Health Organization (WHO) recommends 400 grams of fruits and vegetables per day to reduce the risk of noncommunicable diseases and improve overall health (Wismer, 2009). FAO and WHO introduces the "5-a-day" campaign that encourage people to eat at least five servings of fruits and vegetables daily (FAO, 2006). From 1982 to 1997, per capita consumption of fresh fruits and vegetables in the United States increased from 91.6 to 121.1 kg, an increase of 32% (FDA, 2010). The production of fruits and vegetables has grown steadily over the years with a total global production of ~1500 million tons, led by China (573 million tons) and India (160 million tons). The increasing global trade allows various food (and potential foodborne hazards) to cross borders. As a result, outbreaks of human diseases associated with the consumption of raw fruits, seeds, and vegetables often occur in developing countries and have become more frequent in developed countries over the past two decades (Beuchat, 1998) due to global sourcing.

3.2 Main Risks Associated with Fresh Produce

An apple a day is said to keep doctors at bay—but it is less certain these days, it seems. Raw fruits and vegetables are good but may also send one to the doctor (Noah, 2009). In this chapter we shall look at the main risks associated with fresh produce, contributing factors of on-farm contamination, control measures or preventive measures, and a case study of *E. coli* O157:H7 in bagged spinach.

Fresh produce and sprouted seeds have been implicated in a number of documented outbreaks of illness in countries such as the United States and within the European Union. According to the Center for Disease Prevention and Control (CDC), the number of produce-associated outbreaks per year in the United States doubled between the periods 1973–1987 and 1988–1992 (Olsen et al., 2000). Meanwhile, in the UK, Adak et al. (2005) estimated that between 1996 and 2000, there were 1,723,315 cases of indigenous foodborne disease per year resulting in 21,997 hospitalizations and 687 deaths, but only 3% of cases were attributed to produce. Although fresh produce contributes only 3% of the total cases of foodborne illness in the United Kingdom, this class of food was deemed of particular concern to regulatory authorities (Monaghan et al., 2008). Some fruits and vegetables are likely to be consumed raw or after minimal processing as previously described, and without cooking; so any human pathogens present in the products are given a maximized chance to cause foodborne illness since fruits and vegetables are mostly produced in a natural environment and are vulnerable to contamination (Matthews, 2006; Monaghan et al., 2008). Although most high-profile fresh produce outbreaks have occurred outside the United Kingdom, the fact that they have occurred at all indicates that the controls in place within the industry globally may not have adequately controlling the hazards associated with farming and distribution of fresh produce (Monaghan et al., 2008).

Tauxe et al. (2010) revealed that fresh produce, which have perhaps long been considered among the safest of foods, have now become recognized in the United States as the leading vehicle of illnesses associated with recent foodborne outbreaks. The CDC reported that the incidence of outbreaks is greater for vegetables than for fruits (CDC, 2006), and revealed that salad greens, lettuce, sprouts, melons, and tomatoes are the leading vehicles of illness. These fresh products have received much attention by the FAO/WHO as well, which gave leafy green vegetables (including fresh herbs) the highest priority as commodities of global concern. This ranking was due to the large volume of production and export, the fact that this product type has been associated with numerous outbreaks with varying agents and that the growing and processing stages are becoming more highly complex. A level two priority was given to berries, green onions, melons, sprouted seeds and tomatoes; level three priority was given to a large group comprised of carrots, cucumbers, almonds, baby corn, sesame seeds, onion and garlic, mango, paw paw, and celery (FAO/WHO, 2008). Many of these commodities are vulnerable to contamination because they grow on or close to soil where contamination can occur.

In recent years, the major U.S. produce-associated outbreaks were caused by *Salmonella* and *Escherichia coli* O157:H7. In 2004, three outbreaks and 550 cases were linked to *Salmonella* in tomatoes followed by an *E. coli* outbreak in 2006 due to lettuce consumption at Taco Bell restaurants (CDC, 2006; Corby et al., 2005). While in September 2006, an *E. coli* O157:H7 outbreak linked to the consumption of bagged spinach affected twenty-six states, and involved 205 illnesses and three deaths. Using production codes from the implicated bagged spinach product and by employing DNA fingerprinting, investigators were able to match the *E. coli* O157:H7 isolates from the contaminated spinach to the environmental strains from one field. As there are many potential vehicles for contamination including animals, humans, and water, the exact means by which the spinach contamination occurred remains unknown. The potential environmental risk factors for *E. coli* O157:H7 contamination at or near the field included the presence of wild pigs, the proximity of irrigation wells used to grow produce, and surface waterways exposed to feces from cattle and wildlife (FDA, 2007).

Most of these incidences were caused by fecal contaminated irrigation water and from infected farm workers (Cotterelle et al., 2005; Denny et al., 2007; Emberland et al., 2007; Falkenhorst et al., 2005; Heier et al., 2009; Lofdahl et al., 2009; Pezzoli et al., 2007; Soderstrom et al., 2008; Werner et al., 2007). According to Warriner et al. (2009), foodborne illness outbreaks linked to the fresh-cut chain were specifically due to the inability to control the dissemination of human (and animal) pathogens within the environment; failure of pre- and postharvest interventions to remove field-acquired contamination, and inability to trace contaminated produce back to source. The foodborne pathogens that are frequently associated with fresh produce originate, for the most part, from enteric environments; that is, they are found in the intestinal tract and fecal material of humans or animals (FDA, 2010). Indeed, the fecal–oral route of contamination is a key factor that needs to be controlled within the supply chain. Other pathogens such as *Clostridium botulinum,* which is usually isolated from soils, water, and decaying plant or animal material, and *Listeria monocytogenes,* which can be readily isolated from human and animal feces, as well as from many other environments including soil, agricultural irrigation sources, decaying plant residue on equipment or bins, cull piles, packing sheds, and food processing facilities also need to be adequately controlled (FDA, 2010).

Johnston et al. (2006) stated that the manner in which fruits and vegetables are handled at harvest and postharvest affect their microbiological quality. This is significant because the microbial load on produce at the time of harvest may be carried over to the final product at consumption if no effective intervention processes are in place. Increased consumption of fresh produce, better surveillance, and detection of foodborne outbreaks are likely contributing factors to the increased recognition of fresh produce as vehicles of illnesses.

3.3 Sources of Produce Contamination at the Farm Level

Produce can become contaminated with microbial pathogens by a wide variety of mechanisms. Contamination leading to foodborne illness has occurred during production, harvest, processing, and transporting, as well as in retail and food service establishments and in the home kitchen (FDA, 2010). However, this chapter focuses on preharvest contamination.

3.3.1 Preharvest Practices

The likelihood of the edible parts of a crop becoming contaminated depends upon a number of factors, which includes growing location, type of irrigation application, and nature of produce surface (Table 3.1). Some of the sources of preharvest contamination of produce include irrigation water (Steele and Odumeru, 2004), contaminated manure, sewage sludge, run-off water from livestock operations, and wild and domestic animals (Beuchat, 2006; Delaquis et al., 2007). One of the main safety concerns is the quality of irrigation water. Irrigation techniques that submit plants to direct contact with contaminated water through spraying increases the risk of

TABLE 3.1

Factors Affecting Produce Surface
Contamination during Irrigation

Growing location of the edible portion of the plant
• distance from soil or water surface
Frequency of irrigation
• number of days last irrigated before harvest
Surface of the edible portion
• Smooth
• Webbed
• Rough
Type of irrigation method
• Furrow or flood
• Sprinkler
• Drip—surface/subsurface

Source: Gerba, C. P. and C. Y. Choi. 2009. *The Produce Contamination Problem: Causes and Solutions*, ed. G. M. Sapers, E. B. Solomon, and K. R. Matthews, 105–118. MA, USA: Academic Press.

product contamination as compared to drip irrigation (da Cruz et al., 2006). Irrespective of the application technique, water quality should be considered in all the operations in which it is involved, e.g., irrigation, cooling, pesticide, and fertilizer application. The source of water should be periodically analyzed (Howard and Gonzalez, 2001).

Many produce-associated outbreaks are also linked to contamination with animal manure (CDC, 1999, 2002, 2007; Greene et al., 2008). Contaminated manure can contact the produce directly through its use as soil fertilizer or indirectly through irrigation water or water used during postharvest treatments (Doyle, 2000). Zoonotic agents can also be transferred to the edible portion of crops due to bioaerosol during slurry spreading, rain, or an irrigation event (Hutchinson et al., 2008). Bioaerosol or biological aerosols are biological particles, including pathogenic microorganisms, which have become aerosolized through either human activity such as the land application of biosolids, or through natural activities such as the dispersion of fungal spores (Brooks et al., 2005). Although zoonotic agents on crop surface perish quickly, the risks of overhead irrigation of fresh produce 3 weeks before harvest should still be considered (Hutchinson et al., 2008). This is demonstrated in a study by Solomon et al. (2003), where lettuce plants were spray irrigated with water containing low levels of *E. coli* O157:H7 remain contaminated for extended periods. Hence, efforts to minimize microbial contamination of preharvest crops through surface irrigation to reduce the likelihood of crop contamination (Solomon et al., 2002) or irrigation with potable water if using spray irrigation before harvest (Rangarajan et al., 2000) should be carried out.

Another potential source of pathogenic bacteria in the agricultural environment, which may affect the water used in washing and in pack-houses, is the presence of domestic or wild animals (De Roever, 1998). Birds are an important source of contamination due to their ability to transmit bacteria at great distances. Wild birds are most likely to be involved in contamination while the crops are in the field, and perhaps at nearby field site processing and storage facilities (Kruse et al., 2004). In general, pests and other mammals may often possess enteric pathogens and are therefore potential sources of contamination both in the direct form in the field and by way of the irrigation water (da Cruz et al., 2006). Wild animals have been implicated as possible sources of contamination in fresh produce (Cody et al., 1999; Greene et al., 2008; Jay et al., 2007). Animals that had been demonstrated to carry *E. coli* O157:H7 include, but are not limited to, tame and wild deer (Fischer et al., 2001; Renter et al., 2001; Sargeant et al., 1999), rats (Cizek et al., 1999; Nielsen et al., 2004) and raccoons (Shere et al., 1998). Birds are another singular and important transmission source of pathogens and some of the birds found to be positive for *E. coli* O157:H7 were pigeons (Shere et al., 1998), gulls (Wallace et al., 1997) and starlings (Nielsen et al., 2004). A study by Scaife et al. (2006) in Norfolk, UK, found 20.62% of the fecal samples collected from wild rabbits were positive for VTEC *E. coli* O157.

The potential role of workers in contamination of fresh produce was highlighted in the outbreaks linked to viruses in berries (Herwaldt et al., 1997) and green onions (CDC, 2003; Wheeler et al., 2005). Workers' hygiene is affected by the availability and accessibility of wash water and toilet facilities on the farms and appropriate training of staff. Another issue is the presence of sick workers in the fields (FAO/WHO, 2008), which highlights the need for mandatory health screening. Some significant sources of pathogenic microorganisms on fresh produce both pre- and postharvest and the conditions that influence their survival and growth have been identified (Table 3.2).

TABLE 3.2

Sources of Pathogenic Microorganisms on Fresh Produce and
Conditions that Influence Their Survival and Growth

Preplanting
Source of seeds

Preharvest
Soil
Irrigation water
Green or inadequately composted manure
Air/bioaerosol
Wild and domestic animals
Human handling
Water for other uses (e.g., dilution of pesticides and fertilizers, foliar treatments)

Postharvest
Human handling (workers, consumers)
Harvesting equipment
Transport containers (field to packing shed)
Wild and domestic animals
Air/bioaerosol
Wash and rinse water
Sorting, packing, cutting and further-processing equipment
Ice
Transport vehicles
Improper storage (temperature, physical environment)
Improper packaging (includes new packaging technologies)
Cross-contamination (other foods in storage, preparation, and display areas)
Improper display temperature
Improper handling after wholesale or retail purchase
Cooling water (e.g., hydrocooling)

Source: Adapted from FDA. 2010. Chapter IV. Outbreaks Associated with Fresh and Fresh-Cut Produce. Incidence, Growth, and Survival of Pathogens in Fresh and Fresh-Cut Produce. http://www.fda.gov/Food/Science Research/ResearchAreas/SafePracticesforFoodProcesses/ucm091265. htm (accessed March 15, 2010).

3.4 Prevention and Intervention Strategies

What are the steps that could be taken to prevent or reduce risks of cross-contamination? The following can be taken by growers and are arranged according to inputs (e.g., seeds/transplants, irrigation water, manure or fertilizers) and process (e.g., planting, irrigating, application of manure/fertilizer).

In the European Union, the primary producers are required to follow good practices and manage their operations according to the Annex 1 of Regulation 852/2004. The Regulation states that the application of hazard analysis and critical control point (HACCP) principles to primary production is not yet generally feasible but food hazards present at the primary level of the primary production should be identified and adequately controlled (EU, 2004). This means that HACCP-based procedures should be implemented (Johannessen and Cudjoe, 2009). However, the regulation does not apply to small quantities of primary products. Since 2005, all farmers receiving direct payments from the European Union are subject to compulsory cross-compliance (EC, 2010a). This means that those who receive direct payment are obliged to keep land in good agricultural and environmental condition. Since the protection of soil and land is imperative, this poses as an additional factor for the production of safe fruits and vegetables; that is, by complying with the agricultural and environmental legislation, the pollution and contamination of soil and water will be reduced; hence, cleaner water may be used for irrigation (EC, 2010b). By respecting cross-compliance (agricultural legislation) and implementing and maintaining HACCP-based procedures (food legislation), complemented by national legislation in the respective fields, a sound basis for the production of safe fresh fruits and vegetables is laid (Johannessen and Cudjoe, 2009).

3.4.1 Pest Control, Feral Animal and Domestic Animal Control

With the exception of the *E. coli* O157:H7 outbreak from bagged spinach in 2006, there is no solid evidence of wild animal feces causing produce-associated outbreaks. Feces from wild pigs or cattle were one of the potential causes of the contamination in the spinach. But it is important for farms and members of the public to be aware that wild animals can also act as vectors for infection via the fecal–oral route. Contamination of fresh produce with fecal matter in the fields can be mitigated by having a crop manager or supervisor who can inspect the particular crop before harvesting. If fecal matter were found, the crop manager/supervisor should not harvest the produce that has come into direct contact with the fecal material. However, it is also up to the crop manager's experience and judgment to gauge the safe distance for harvesting between the fecal matter and other fresh produce. California Leafy Greens Marketing Agreement recommended a minimum 5-foot radius buffer distance (LGMA, 2010).

3.4.2 Manure Management

Some reports demonstrated that pathogens like *E. coli* O157, *Salmonella enterica,* and *Listeria monocytogenes* are able to survive for extended periods (up to 3 months) in manure (Franz et al., 2005; Scott et al., 2006) and manure-amended soil (Franz et al., 2008). Aging or proper manure management is a heat pasteurization process at (140–149°F/60–65°C) (Kudva et al., 1998). Manure composting refers to controlled aerobic and thermophilic (131–149°F/55–65°C) decomposition of organic matter by microorganisms. Manure must be aerated and allowed to reach peak microbial composition in at least 3 months (Millner, 2009; Suslow et al., 2003). Self-heating of stacked manure without attention to the time and temperature needed to reduce pathogen load does not adequately meet the composting requirement (Millner, 2009).

3.4.3 Water

Irrigation water is a potential point of pathogen entry into the food chain as many bacteria, viruses, and protozoa of fecal origin can be found in waters which are used in the primary production of food crops (Ethelberg et al., 2010; Falkenhorst et al., 2005; Lofdahl et al., 2009; Nygard et al., 2008; Soderstrom et al., 2008). Pathogens can be taken up on plant surfaces especially at the point of harvesting and trimming wounds, damage caused during handling, and processing or natural points of entry (Suslow et al., 2003). The predominant irrigation method (99%) in the United Kingdom for salad crops is overhead irrigation (Tyrrel et al., 2006). Overhead irrigation is often highlighted as carrying a higher risk for ready-to-eat crops than subsurface or drip systems since spray irrigation exposes the edible parts of crop directly to water. Crops with rough surfaces or leaves may also retain more water (Gerba and Choi, 2009). The application of irrigation water is usually planned to ensure optimum crop quality and growth. The irrigation schedule also varies, depending on crop types, crop water demand, soil, rainfall and weather parameters. A longer harvest interval may affect the number of pathogens on crop surfaces since harvest interval will allow the effects of UV radiation, temperature, and desiccation to reduce the pathogen load (Groves et al., 2002).

Besides irrigation water, the use of contaminated water for washing, hydrocooling, and icing may also lead to produce contamination. According to Gagliardi et al. (2003), a hydrocooler was found to be a source of fecal coliforms and fecal enterococci contamination in melon rinds. Suslow et al. (2003) suggested that when microbes are taken up on plant surfaces they are then unaffected by postharvest water treatments with chlorine, chlorine dioxide, ozone, peroxide, or peroxyacetic acid. They proposed that there must be sufficient disinfection of the water before it comes into contact with water during washing, cooling, and transportation using flotation tanks or flumes or drenching, otherwise, there is the potential for microorganisms to attach or penetrate the produce. Suslow et al. (2003) proposed that the microbiological standards for water should become more restrictive the closer the produce gets to the proposed harvest date, and in postharvest handling facilities.

Crops with rough surfaces or leaves may also retain more water (Gerba and Choi, 2009). The application of irrigation water is usually planned to ensure optimum crop quality and growth. The irrigation schedule also varies, depending on crop types, crop water demand, soil, rainfall, and weather parameters. A longer harvest interval may affect the number of pathogens on crop surfaces since harvest interval will allow the effects of UV radiation, temperature, and desiccation to reduce the pathogen load (Groves et al., 2002). Primary factors affecting the survival of pathogens in irrigation water and on the crop are temperature, sunlight, storage time of water, pH, harvest interval, and protection afforded by the crop itself, e.g., multiple leaf structures (FSA, 2010). The report further determined that as sunlight and temperature are primary factors for inactivation of pathogens, crops with a long harvest interval between the last episodes of irrigation and when the crop is harvested gives a greater opportunity for pathogens to die. Shallow rooting crops such as spinach and leafy salad, or crops grown in protected environments (greenhouses or tunnels) require more regular irrigation and thus have the potential to be at increased risk especially as they are ready-to-eat crops (RTE). Monaghan and Hutchison (2010) reviewed the standards for selected indicator numbers in irrigation water for crops that are likely to be eaten uncooked (Table 3.3). They determined that in the United Kingdom there are no statutory criteria for indicator bacteria in irrigation water, rather a range of international standards.

3.4.4 Workers' Hygiene

The health and hygiene of all workers who handle fresh produce, whether it's on a farm, in the packinghouse, markets, grocery store, or food service establishment, are of significant importance in preventing produce-associated outbreaks. Organisms such as Hepatitis A virus (Wheeler et al., 2005) and norovirus (Falkenhorst et al., 2005) had been implicated to spread to produce via the fecal–oral route of transmission from infected workers who work when they were ill. Farm workers, especially seasonal workers, tend to be a transitory group, so farms may have someone new every year. Similarly, farm workers are not required to be vaccinated, which presents an increased threat for the spread of disease, particularly among foods that do not require cooking (*Science Daily*, 2009). According to Gravani (2009), one of the best strategies for preventing contamination by workers is a well-designed and well-delivered education and training program. Clayton et al. (2003) highlighted that training needs to be based around a risk-based approach and that behavioral change will not occur merely as a result of training. The concept of risk is considered an important part of food hygiene training. Clayton

TABLE 3.3

International Standards and Guidelines for Selected Indicator Numbers in
Irrigation Water for Crops that are Likely to be Eaten Uncooked

Issuing Body	Indicator Bacteria	Performance Criteria
World Health Organization Treated wastewater	Fecal coliforms	≤ 1000 CFU/100 ml (calculated as a geometric mean)
State of California, recycled irrigation water	Total coliforms	≤2.2 MPN CFU/100 ml in previous 7 days of test results. No sample to exceed 23 MPN CFU/100 ml in previous month
Canadian Agriculture Ministry, irrigation water	Fecal coliforms or *E. coli* and also Total coliforms	≤100 CFU of fecal coliforms or *E. coli* per 100 ml
Tesco Stores Nurture Scheme. Irrigation water	*E. coli* and also Fecal coliforms	≤1,000 CFU of total coliforms per 100 m
		≤1,000 CFU/100 ml for both indicators (calculated as a geometric mean if multiple samples are taken)
Marks & Spencer Field to Fork. Irrigation water	*E. coli*	≤1,000 CFU/100 m

Source: Monaghan, J. and M. Hutchinson. 2010. Monitoring Microbial Food Safety of Fresh Produce. Factsheet 13/10 Edible Crops. Jointly produced by Horticultural Development Company and Food Standards Agency, available at: http://www.food.gov.uk/multi-media/pdfs/microbial.pdf (accessed January 3, 2012). With permission.

Note: CFU, colony forming units; MPN, a test result calculated from the most probable number microbiological test method.

et al. (2002) found out that an increase in the knowledge of a food handler does not necessarily change their food handling behavior, and it depends upon their attitude. Hence, consideration should also be given to cultural and social aspects, including longstanding entrenched worker behaviors, attitudes, and social taboos (FAO/WHO, 2008).

3.4.5 Produce Sanitization

The harvesting and processing of fresh produce involves significant handling by agricultural workers. Humans represent a significant source of pathogens that can be readily transferred to produce and subsequently to consumers (NACMCF, 1999). The most significant risk associated with handling is the possible introduction of enteric viruses (Koopmans and Duizer, 2004; Koopmans et al., 2003). The potential role of workers in contamination of fresh produce was highlighted in the outbreaks linked to viruses in berries and green onions. Workers' hygiene is affected by the availability and accessibility of wash and toilet facilities on the farms and appropriate training of staff. Another issue is the presence of sick workers in the fields

(FAO/WHO, 2008), which highlighted that there should be health screening in place. Sources of pathogenic microorganisms on fresh produce both pre- and postharvest and the conditions that influence their survival and growth have been identified.

These potential sources of microbial pathogens need to be adequately controlled in order to manage the level of incidence of foodborne illness outbreaks. Maybe a banana a day would be better then ... provided you peel it yourself first, of course.

3.5 Case Study of Breakdown in Fresh Produce Farms

At the time of writing, we had considered using the *E. coli* O104:H4 outbreak in Germany, which occurred from May–July 2011. This will make a fascinating case study when it comes to outbreak investigation, procedures used in identifying this unique strain, and discussions on why it strangely affects more women and adults. The outbreak has also caused the highest number of hemolytic uremic syndrome in history. Finally, fenugreek seeds were found as the most likely link to the outbreak in Germany and France, and the source were traced back to Egypt. European Food Safety Authority (EFSA) extended the ban of sprouted seed till 31 March 2012 due to the investigation for source of contamination by the Egyptian counterparts. The year 2011 was also a significant year for Egypt due to the political turmoil occurrences. Hence, we decided to focus on previous fresh produce outbreaks that had been confirmed to link to on-farm contamination. In this case, we will be able to identify weak point sources of contamination and apply intervention strategies to reduce or prevent the same mistakes from occurring.

3.5.1 *Escherichia Coli* O157:H7 in Bagged Spinach, United States, 2006

In August/September 2006, an *E. coli* O157:H7 outbreak linked to the consumption of bagged spinach affected 26 U.S. states, and involved 205 illnesses and three deaths. Using production codes from the implicated bagged spinach product and by employing pulsed-field gel electrophoresis (PFGE), investigators were able to match the *E. coli* O157:H7 isolates from the contaminated spinach to the environmental strains from one field. Contaminated product was traced to one production date (August 15, 2006) at one processing plant and fields located on four ranches on the central California coast (Figure 3.1) (CALFERT, 2007). As there are many potential vehicles for contamination including animals, humans, and water, the exact means by which the spinach contamination occurred remains unknown. The potential environmental risk factors for *E. coli* O157:H7 contamination at or near the field included the presence of wild pigs, the proximity of irrigation wells used to grow produce, and surface waterways exposed to feces from cattle and wildlife (FDA, 2007). (See Figure 3.1.)

The outbreak strain was isolated from cattle feces collected about 1 mile from an implicated spinach field on a ranch. A number of free-roaming feral swine were observed there. Hence, Jay et al. (2007) investigated the potential involvement of feral swine in *E. coli* O157 contamination of spinach fields. Fecal and swab samples were collected and molecular typing of isolates by pulsed-field gel electrophoresis (PFGE) and multiple-locus variable number (MLVA) were conducted. The percentage of specimens positive for *E. coli*

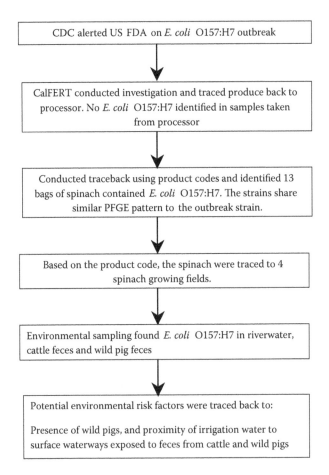

FIGURE 3.1
Traceback investigation of *Escherichia coli* O157: H7 in bagged spinach. (Adapted from CALFERT.2007. Investigation of an *Escherichia coli* O157:H7 Outbreak Associated with Dole Pre-Packaged Spinach. Final March 21, 2007. California Department of Health Services, Sacramento, California and the US Food and Drug Administration, Almeda, California.http:// www.marlerclark.com/2006_Spinach_Report_Final_01.pdf [accessed December 14, 2009].)

O157 among feral swine was 14.9% and cattle 33.8%. There was a relatively high density of feral swine near the cattle and spinach fields (4.6 swine/km²). PFGE pattern and MLVA analyses revealed possible swine-to-swine transmission, interspecies transmission between cattle and swine, or through a common source of exposure such as water or soil. Cattle and feral swine had access to and congregated at surface waterways on the ranch (Figure 3.2i, Figure 3.3). Feral swine used livestock rangelands and gained access to adjacent crop fields through gaps formed at the base of the fence by erosion and rooting (Figure 3.2iv) (CALFERT, 2007).

(i)

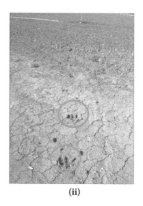

(ii)

(iii)

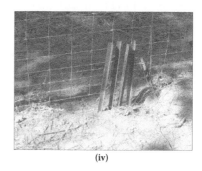

(iv)

FIGURE 3.2

(Clockwise from top left) (i) Cattle crossing river near the spinach fields; (ii) pig tracks found on dirt road heading into nearby implicated farm; (iii) pig rooting and tracks found in neighboring spinach fields; and (iv) hole underneath fence in implicated ranch. (Adapted from: CALFERT.2007. Investigation of an *Escherichia coli* O157:H7 outbreak associated with Dole pre-packaged spinach. Final March 21, 2007. California Department of Health Services, Sacramento, California and the US Food and Drug Administration, Almeda, California.http://www.marlerclark.com/2006_Spinach_Report_Final_01.pdf (accessed December 14, 2009). With permission.)

3.5.2 Investigation of Irrigation Water as a Potential Source of Outbreak

The California Food Emergency Response Team (CALFERT), U.S. Food and Drug Administration (FDA), California Department of Public Health Food and Drug Branch (FDB), and the Center for Disease Control and Prevention (CDC) collaborated to carry out the environmental investigation of the outbreak. Water sources for irrigation were identified and relevant data such as:

- Drillers' logs for wells on and near farms identified by traceback investigations
- Locations of irrigation wells relative to contamination sources and surface waters
- Records of depths to groundwater in monitoring wells over time
- Water quality analyses for both groundwater wells and surface waters
- Data on location and timing of percolation from surface waters into groundwater
- Information on flow in rivers and streams, and
- Records of direct use of surface water for irrigation on farms

The most probable environmental risk factors were the presence of wild pigs and fecal contaminated irrigation water from cattle and pigs. Although the farm used groundwater water for irrigation, there is a possibility that the groundwater were recharged with contaminated surface water when the groundwater level fell below the riverbed level in August/September 2006 (Figure 3.3; Gelting et al., 2011). This also coincides with the onset of outbreak.

3.5.3 Contamination Traced Back to the Farm; Now What?

When contaminations are traced back to the farm, a farm investigation is usually initiated. One should be aware of the origin of the etiologic agent—is it human or animal origin? Common examples are *E. coli* O157:H7—cattle; *Salmonella enteritidis*—poultry; *Salmonella* typhimurium—swine; and *Campylobacter* spp.—poultry. While viruses detected in produce such as Hepatitis A virus are usually of human origin, other viruses such as Hepatitis E were found in both human and swine. Noroviruses were detected in humans (but had been found in calf and pigs, although the strains were genetically distinct from humans) (van Der Poel et al., 2000). Protozoan associated outbreak such as *Cryptosporidium parvum* is associated with humans and cattle, *Cyclospora cayetanensis*—human, and *Giardia* spp.—beavers (Rose and Slifko, 1999).

Some of the steps that can be carried out at the farm level are:

i. Obtain map of farm location and identify sources of irrigation water, wells and surface water, nearby farmland, pastures, and surrounding environment.

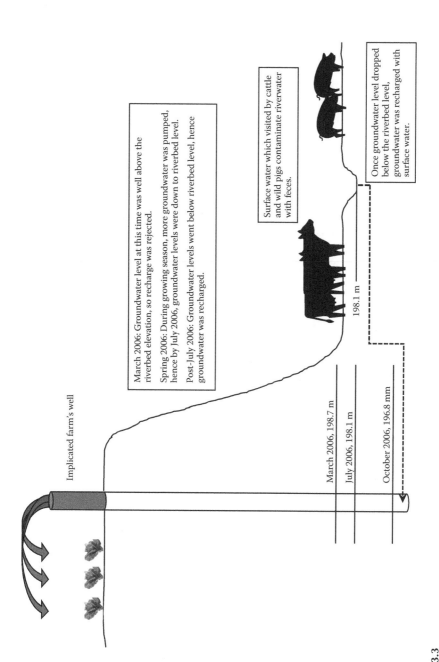

Implicated farm's well

March 2006: Groundwater level at this time was well above the riverbed elevation, so recharge was rejected.

Spring 2006: During growing season, more groundwater was pumped, hence by July 2006, groundwater levels were down to riverbed level.

Post-July 2006: Groundwater levels went below riverbed level, hence groundwater was recharged.

Surface water which visited by cattle and wild pigs contaminate riverwater with feces.

Once groundwater level dropped below the riverbed level, groundwater was recharged with surface water.

198.1 m

March 2006, 198.7 m

July 2006, 198.1 m

October 2006, 196.8 mm

FIGURE 3.3
Potential contamination of spinach from fecally contaminated irrigation water. (Adapted from Gelting, R. J. et al. 2011. *Agr. Water Manage.* 98: 1395–1402.)

ii. Observe for presence (or evidence) of domestic or wild animals within farming area or farm surroundings.

iii. Collect environmental samples (soil, irrigation water, wild/domestic animal feces).

iv. If pathogens are suspected to be human origin, conduct interviews among farm workers and collect fecal samples.

v. Review map, farm inventory, records and dates of inputs (seedlings/plants, irrigation water, fertilizers/compost, pesticides, source of water used for diluting pesticides and fertilizers, application method), and how and when the crops are harvested and transported (Farrar and Guzewich, 2009).

An important tool in outbreak and environmental investigations is DNA fingerprinting of strains to identify links and potential sources of contamination (Cooley et al., 2007). PFGE is instrumental in linking geographically dispersed *E. coli* O1057:H7 isolates recovered from the stool samples of ill persons, and linking these human isolated to isolates from spinach (Wendel et al., 2009).

References

Abadias, M., J. Usall, M. Anguera, C. Solsona, and I. Viñas. 2008. Microbiological quality of fresh, minimally-processed fruit and vegetables, and sprouts from retail establishments. *Int. J. Food Microbiol.* 123: 121–129.

Adak, G. L., S. M. Meakins, H. Yip, B. A. Lopman, and S. J. O'Brien. 2005. Disease risks from foods, England and Wales, 1996–2000. *Emerg. Infect. Dis.* 11: 365–372.

Beuchat, L. R. 1998. Surface Decontamination of Fruits and Vegetables Eaten Raw: A Review. Food Safety Unit, World Health Organization, WHO/FSF/98.2. http://www.who.int/foodsafety/publications/fs_management/en/surface_decon.pdf (accessed January 11, 2010).

Beuchat, L. R. 2006. Vectors and conditions for preharvest contamination of fruits and vegetables with pathogens capable of causing enteric diseases. *Brit. Food J.* 108: 38–53.

Bhagwat, A. 2006. Microbiological safety of fresh produce: Where are we now? In *Microbiology of Fresh Produce*, ed. K. R. Matthews, 121–165. Washington, D.C. ASM Press.

Brooks, J. P., B. D. Tanner, C. P. Gerba, C. N. Haas, and I. L. Pepper. 2005. Estimation of bioaerosol risk of infection to residents adjacent to a land applied biosolids site using an empirically derived transport model. *J. Appl. Microbiol.* 98: 397–405.

CALFERT. 2007. Investigation of an *Escherichia coli* O157:H7 Outbreak Associated with Dole Pre-Packaged Spinach. Final March 21, 2007. California Department of Health Services, Sacramento, California and the US Food and Drug Administration, Almeda, California.http://www.marlerclark.com/2006_Spinach_Report_Final_01.pdf (accessed December 14, 2009).

CDC. 1999. Outbreaks of *Shigella sonnei* infection associated with eating fresh parsley—United States and Canada, July–August 1998. *MMWR Morbidity Mortality Weekly Report* 48: 285–289.

CDC. 2002. Multistate outbreaks of *Salmonella* serotype Poona infections associated with eating cantaloupe from Mexico–United States and Canada, 2000–2002. *MMWR Morbidity Mortality Weekly Report* 51: 1044–1047.

CDC. 2003. Hepatitis A outbreak associated with green onions at a restaurant—Monaca, Pennsylvania, 2003. *MMWR Morbidity Mortality Weekly Report* 52: 1155–1157.

CDC. 2006. Multistate outbreak of *E. coli* O157 infections, November–December 2006. *MMWR Morbidity Mortality Weekly Report* 55: 1045–1046.

CDC. 2007. Multistate outbreaks of *Salmonella* infections associated with raw tomatoes eaten in restaurants—United States, 2005–2006. *MMWR Morbidity Mortality Weekly Report* 56: 909–911.

Cizek, A., P. Alexa, I. Literak, J. Hamrik, P. Novak, and J. Smola. 1999. Shiga toxin-producing *Escherichia coli* O157 in feedlot cattle and Norwegian rats from a large-scale farm. *Lett. Appl. Microbiol.* 28: 435–439.

Clayton, D. A., C. J. Griffith, P. Price, and A. C. Peters. 2002. Food handler's beliefs and self-reported practices. *Int. J. Environ. Heal. R.* 12: 25–39.

Clayton, D. A., C. J. Griffith, and P. Price. 2003. An investigation of the factors underlying consumers' implementation of specific food safety practices. *Brit. Food J.* 105: 434–453.

Cody, S. H., M. K. Glynn, J. A. Farrar et al. 1999. An outbreak of *Escherichia coli* O157:H7 infection from unpasteurized commercial apple juice. *Ann. Intern. Med.* 133: 202–209.

Cooley, M., D. Carychao, L. Crawford-Miksza et al. 2007. Incidence and tracking of *Escherichia coli* O157:H7 in a major produce production region in California. *PloS ONE* 2: e1159.

Corby, R., V. Lanni, V. Kitsler et al. 2005. Outbreaks of *Salmonella* infections associated with eating tomatoes—United States and Canada, 2004. *Morbidity and Mortality Weekly Report* 54: 325–328.

Cotterelle, B., C. Drougard, J. Rolland et al. 2005. Outbreak of norovirus infection associated with the consumption of frozen raspberries, France, March 2005. *Euro Surveill* 10: pii=2690. http://www.eurosurveillance.org/ViewArticle.aspx?ArticleId=2690 (accessed January 1, 2010).

da Cruz, A. G., S. A. Cenci, and M. C. Antun Maia. 2006. Good agricultural practices in a Brazilian produce plant. *Food Control* 9: 321–347.

Delaquis, P., S. Bach, and L.-D. Dinu. 2007. Behavior of *Escherichia coli* O157:H7 in leafy vegetables. *J. Food Protect.* 70: 1966–1974.

Denny, J., J. Threlfall, J. Takkinen et al. 2007. Multinational *Salmonella* Paratyphi B variant Java (*Salmonella* Java) outbreak, August–December 2007. *Euro Surveill* 12: pii=3332. http://www.eurosurveillance.org/ViewArticle.aspx?ArticleId=3332 (accessed December 31, 2009).

De Roever, C. 1998. Microbiological safety evaluations and recommendations on fresh produce. *Food Control* 9: 321–347.

Doyle, M. P. 2000. Food safety issues arising at food production in a global market. *J. Agribusiness* 18: 129–133.

EC. 2010a. Agriculture and the Environment: Cross Compliance. http://ec.europa.eu/agriculture/envir/cross-compliance/index_en.htm (accessed September 11, 2011).

EC. 2010b. Agriculture and the Environment: Water. http://ec.europa.eu/agriculture/envir/water/index_en.htm (accessed September 11, 2011).

Emberland, K. E., S. Ethelberg, M. Kuusi et al. 2007. Outbreak of *Salmonella* Weltevreden infections in Norway, Denmark and Finland Associated with Alfalfa Sprouts, July–October 2007. *Euro Surveill* 12:pii = 3321. http://www.eurosurveillance.org/ViewArticle.aspx?ArticleId=3321 (accessed December 14, 2009).

Ethelberg, S., M. Lisby, B. Böttiger et al. 2010. Outbreaks of Gastroenteritidis Linked to Lettuce, Denmark, January 2010. *Euro Surveill* 15: pii = 19484. http://www.eurosurveillance.org/ViewArticle.aspx?ArticleId=19484 (accessed February 14, 2010).

EU. 2004. Corrigendum to Regulation (EC) No 852/2004 of the European Parliament and of the Council of 29 April 2004 on the hygiene of foodstuffs (OJ L 139, 30.4.2004). *Off. J. Eur. Union* 226: 3–21. http://eur-lex.europa.eu/LexUriServ/LexUriServ.do?uri=OJ:L:2004:226:0003:0021:EN:PDF (accessed January 17, 2010).

FAO. 2006. More Fruits and Vegetables. http://www.fao.org/ag/magazine/0606sp2.htm (accessed January 2, 2011).

FAO/WHO. 2008. Microbiological Hazards of Fresh Fruits and Vegetables. FAO/WHO, Rome/Geneva: pp. 1–28.

Falkenhorst, G., L. Krusell, M. Lisby, S. Bo Madsen, B. Bottiger, and K. Molbak. 2005. Imported frozen raspberries cause a series of norovirus outbreaks in Denmark, 2005. *Euro Surveill.* 10: pii = 2795. http://www.eurosurveillance.org/ViewArticle.aspx?ArticleId=2795 (accessed December 15, 2009).

Farrar, J. and J. Guzewich. 2009. Identification of the source of contamination. In *The Produce Contamination Problem: Causes and Solutions*, ed. G. M. Sapers, E. B. Solomon, and K. R. Matthews, 49–77. Burlington, MA: Academic Press.

FDA. 2007. FDA finalizes report on 2006 Spinach Outbreak. http://www.fda.gov/NewsEvents/Newsroom/PressAnnouncements/2007/ucm108873.htm (accessed December 23, 2009).

FDA. 2010. Chapter IV. Outbreaks Associated with Fresh and Fresh-Cut Produce. Incidence, Growth, and Survival of Pathogens in Fresh and Fresh-Cut Produce. http://www.fda.gov/Food/ScienceResearch/ResearchAreas/SafePracticesforFoodProcesses/ucm091265.htm (accessed March 15, 2010).

Fischer, J. R., T. Zhao, M. Doyle et al. 2001. Experimental and field studies of *Escherichia coli* O157:H7 in white-tailed deer. *Appl. Environ. Microbiol.* 67: 1218–1224.

Franz, E., A. D. van Diepeningen, O. J. de Vos, and A. H. C. van Bruggen. 2005. Effects of cattle feeding regimen and soil management type on the fate of *Escherichia coli* O157:H7 and *Salmonella enterica* serovar Typhimurium in manure, manure-amended soil, and lettuce. *Appl. Environ. Microbiol.* 71: 6165–6174.

Franz, E., A. V. Semenov, A. J. Termorshuizen, O. J. de Vos, J. G. Bokhorst, and A. H. C. van Bruggen. 2008. Manure-amended soil characteristics affecting the survival of *E. coli* O157:H7 in 36 Dutch soils. *Environ. Microbiol.* 10: 313–327.

FSA. 2010. Agency Helping UK Growers Keep Fresh Produce Clean and Safe. http://www.food.gov.uk/news/newsarchive/2010/sep/growers (accessed November 20, 2010).

Gagliardi, J. V., P. D. Millner, G. Lester, and D. Ingram. 2003. On-farm and post-harvest processing sources of bacterial contamination to melon rinds. *J. Food Protect.* 66: 82–87.

Gelting, R. J., M. A. Baloch, M. A. Zarate-Bermudez, and C. Selman. 2011. Irrigation water issues potentially related to the 2006 multistate *E. coli* O157:H7 outbreak associated with spinach. *Agr. Water Manage.* 98: 1395–1402.

Gerba, C. P. and C. Y. Choi. 2009. Water quality. In *The Produce Contamination Problem: Causes and Solutions,* ed. G. M. Sapers, E. B. Solomon, and K. R. Matthews, 105–118. MA, USA: Academic Press.

Gravani, R. 2009. The role of good agricultural practices in produce safety. In *Microbial Safety of Fresh Produce*, ed. X. Fan, B. A. Niemira, C. J. Doona, F. E. Feeherry, and R. B. Gravani, 101–117. Iowa: Wiley-Blackwell.

Greene, S. K., E. R. Daly, E. A. Talbot et al. 2008. Recurrent multistate outbreak of *Salmonella* Newport associated with tomatoes from contaminated fields, 2005. *Epidemiol Infect* 136: 157–165.

Groves, S. J., N. Davies, and M. N. Aitken. 2002. A Review of the Use of Water in UK Agriculture and the Potential Risks to Food Safety. A Report to the Food Standards Agency, pp. 1–133. http://www.foodbase.org.uk//admintools/reportdocuments/194-1-328_B17001_FSA_Final_Report.pdf (accessed August 8 2010).

Heier, B. T., K. Nygard, G. Kapperud, B. A. Lindstedt, G. S. Johannessen, and H. Blekkan. 2009. *Shigella sonnei* infections in Norway associated with sugar peas, May–June 2009. *Euro Surveill*14:pii=19243. http://www.eurosurveillance.org/ViewArticle.aspx?ArticleId = 19243 (accessed December 31, 2009).

Herwaldt, B. L., M.-L. Ackers, and the Cyclospora Working Group. 1997. An outbreak in 1996 of cyclosporiasis associated with imported raspberries. *N. Engl. J. Med.* 336: 1548–1556.

Howard, L. R. and E. R. Gonzalez. 2001. Food safety and produce operations: What is the future? *Hortscience* 36: 20–39.

Hutchinson, M. L., S. M. Avery, and J. M. Monaghan. 2008. The air-borne distribution of zoonotic agents from livestock waste spreading and microbiological risk to fresh produce from contaminated irrigation sources. *J Appl. Microbiol.* 105: 848–857.

Isshiki, K., Md. Latiful Bari, S. Kawamoto, and T. Shiina. 2009. Regulatory issues in Japan regarding produce safety. In *The Produce Contamination Problem: Causes and Solutions*, ed. G. M. Sapers, E. B. Solomon, and K. R. Matthews, 353–389. MA, USA: Academic Press.

Jay, M. T., M. Cooley, D. Carychao et al. 2007. *Escherichia coli* O157:H7 in feral swine near spinach fields and cattle, Central California Coast. *Emerg Infect Dis* 13: 1908–1911.

Johannessen, G. S. and K. S. Cudjoe. 2009. Regulatory issues in Europe regarding fresh fruit and vegetable safety. *Causes and Solutions*, ed. G. M. Sapers, E. B. Solomon and K. R. Matthews, 331–352. MA, USA: Academic Press.

Johnston, L. M., L.-A.Jaykus, D. Moll, J. Anciso, B. Mora, and C. L. Moe. 2006. A field study of the microbiological quality of fresh produce of domestic and Mexican origin. *Int. J. Food Microbiol.* 112: 83–95.

Koopmans, M., H. Vennema, H. Heersma et al. 2003. Early identification of common-source foodborne virus outbreaks in Europe. *Emerg. Infect. Dis.* 9: 1136–1142.

Koopmans, M. and E. Duizer, E. 2004. Foodborne viruses: An emerging problem. *Int. J. Food Microbiol.* 90: 23–41.

Kruse, H., A. M. Kirkemo, and K. Handeland. 2004. Wildlife as source of zoonotic infections. *Emerg. Infect. Dis.* 10: 2067–2072.

Kudva, I. T., K. Blanch, and C. J. Hovde. 1998. Analysis of *Escherichia coli* O157:H7 survival in ovine or bovine manure and manure slurry. *App. Environ. Microbiol.* 64: 3166–3174.

LGMA. 2010. California Leafy Green Products Handler Marketing Agreement. Commodity Specific Food Safety Guidelines for the Production and Harvest of Lettuce and Leafy Greens. http://www.caleafygreens.ca.gov/members/documents/LGMAAcceptedFoodSafetyPractices01.29.10.pdf (accessed August 13, 2010).

Lofdahl, M., S. Ivarsson, S. Andersson, J. Langmark, and L. Plym-Forshell, L. 2009. An outbreak of *Shigella dysenteriae* in Sweden, May–June 2009, with sugar snaps as the suspected source. *Euro Surveill.* 14:pii = 19268. http://www.eurosurveillance.org/ViewArticle.aspx?ArticleId=19268 (accessed December 31, 2009).

Matthews, K. R. 2006. Microorganisms associated with fruits and vegetables. In *Microbiology of Fresh Produce*, ed. K. R. Matthews, 1–19. Washington, D.C., ASM Press.

Millner, P. D. 2009. Manure management. In *The Produce Contamination Problem: Causes and Solutions*, ed. G. M. Sapers, E. B. Solomon and K. R. Matthews, 79–104. MA, USA: Academic Press.

Monaghan, J. and M. Hutchinson. 2010. Monitoring Microbial Food Safety of Fresh Produce. Factsheet 13/10 Edible Crops. Jointly produced by Horticultural Development Company and Food Standards Agency, available at: http://www.food.gov.uk/multimedia/pdfs/microbial.pdf (accessed January 3, 2012).

Monaghan, J. M., D. J. I. Thomas, K. Goodburn, and M. L. Hutchison. 2008. A review of the published literature describing foodborne illness outbreaks associated with ready to eat fresh produce and an overview of current UK fresh produce farming practices. *Food Standards Agency Project B17007*, pp. 1–222, available at: http://www.foodbase.org.uk//admintools/reportdocuments/340-1-596_B17007_Final_Published_Report.pdf (accessed February 27, 2010).

NACMCF. 1999. Microbiological safety evaluations and recommendations on sprouted seeds. *Int. J. Food Microbiol.* 52: 123–153.

Nielsen, E. M., M. N. Skov, J. J. Madsen, J. Lodal, J. B. Jespersen, and D. L. Baggesen. 2004. Verocytotoxin-producing *Escherichia coli* in wild birds and rodents in close proximity to farms. *Appl. Environ. Microbiol.* 70: 6944–6947.

Noah, N. 2009. Food poisoning from raw fruit and vegetables. *Epidemiol. Infect.* 137: 305–306.

Nygard, K., J. Lassen, L. Vold et al. 2008. Outbreak of *Salmonella* Thompson infections linked to imported rucola lettuce. *Foodborne Pathog. Dis.* 5: 165–173.

Olsen, S., L. Mckinnon, J. Goulding, N. Bean, and L. Slutsker. 2000. Surveillance for foodborne disease outbreaks—United States, 1993–1997. *Mortality and Morbidity Weekly Report* 1: 1–51.

Pezzoli, L., R. Elson, C. Little et al. 2007. International Outbreak of *Salmonella* Sefternberg in 2007. *Euro Surveill.* 12:pii = 3218. http://www.eurosurveillance.org/ViewArticle.aspx?ArticleId = 3218 (accessed December 31, 2009).

Rangarajan, A., E. A. Bihn, R. B. Gravani, D. L. Scott, and M. P. Pritts. 2000. Food Safety Begins on the Farm. A Grower's Guide. Good Agricultural Practices for Fresh Fruits and Vegetables. http://ecommons.cornell.edu/handle/1813/2209?mode=full (accessed January 18, 2011).

Renter, D. G., J. M. Sargeant, S. E. Hygnstorm, J. D. Hoffman, and J. R. Gillespie. 2001. *Escherichia coli* O157:H7 in free-ranging deer in Nebraska. *J. Wildl. Dis.* 37: 755–760.

Rose, J. B. and T. R. Slifko. 1999. *Giardia, Cryptosporidium,* and *Cyclospora* and their impact on foods: A review. *J. Food Prot.* 62: 1059–1070.

Sargeant, J. M., D. J. Hafer, J. R. Gillespie, and R. D. Oberst. 1999. Prevalence of *Escherichia coli* O157:H7 in white-tailed deer sharinf rangeland with cattle. *J. Am. Vet. Med. Assoc.* 215: 792–794.

Scaife, H. R., D. Cowan, J. Finney, S. F. Kinghorn-Perry, and B. Crook. 2006. Wild rabbits (*Oryctolagus cuniculus*) as potential carriers of verocytotoxin-producing *Escherichia coli. Vet. Rec.* 159: 175–178.

Science Daily. 2009. Field of Germs: Food Safety is in Farm Worker's Hands. http://www.sciencedaily.com/releases/2009/02/090220074835.htm (accessed February 22, 2011).

Scott, L., P. McGee, J. J. Sheridan, B. Earley, and N. Leonard. 2006. A comparison of the survival in faeces and water of *Escherichia coli* O157:H7 grown under laboratory conditions or obtained from cattle faeces. *J. Food Protect.* 69: 6–11.

Shere, J. A., K. J. Bartlett, and C. W. Kaspar. 1998. Longitudinal study of *Escherichia coli* O157:H7 dissemination on four dairy farms in Wisconsin. *Appl. Environ. Microbiol.* 64: 1390–1399.

Soderstrom, A., P. Osterberg, A. Lindqvist et al. 2008. A large *Escherichia coli* O157 outbreak in Sweden associated with locally produced lettuce. *Foodborne Pathog. Dis.* 5: 339–349.

Solomon, E. B., H.-J. Pang, and K. R. Matthews. 2003. Persistence of *Escherichia coli* O157:H7 on lettuce plants following spray irrigation with contaminated water. *J. Food Protect.* 66: 2198–2202.

Solomon, E. B., C. J. Potenski, and K. R. Matthews. 2002. Effect of irrigation method on transmission to and persistence of *Escherichia coli* O157:H7 on lettuce. *J. Food Protect.* 65: 673–676.

Steele, M. and J. Odumeru. 2004. Irrigation water as source of foodborne pathogens on fruits and vegetables. *J. Food Protect.* 67: 2839–2849.

Suslow, T.V., M. P. Oria, L. R. Beuchat et al. 2003. Production practices as risk factors in microbial food safety of fresh and fresh-cut produce. *Compr. Rev. Food Sci. Food Saf.* 2: 38–77.

Tauxe, R. V., M. P. Doyle, T. Kuchenmüller, J. Schlundt, and C. E. Stein. 2010. Evolving public health approaches to the global challenge of foodborne infections. *Int. J. Food Microbiol.* 139: S16–S28.

Tyrrel, S. F., J. W. Knox, and E. K. Weatherhead. 2006. Microbiological water quality requirements for salad irrigation in the United Kingdom. *J. Food Protect.* 69: 2029–2035.

Van Der Poel, W. H., J. Vinjé, R. van Der Heide, M. I. Herrera, A. Vivo, and M. P. Koopmans 2000. Norwalk-like calicivirus genes in farm animals. *Emerg. Infect. Dis.* 6: 36–41.

Wallace, J. S., T. Cheasty, and K. Jones. 1997. Isolation of verocytotoxin-producing *Escherichia coli* O157 from wild birds. *J. Appl. Microbiol.* 82: 399–404.

Warriner, K., A. Huber, A. Namvar, Wei Fan, and K. Dunfield. 2009. Chapter 4: Recent advances in the microbial safety of fresh fruits and vegetables. In *Advances in Food and Nutrition Research,* ed. S. L. Taylor, Vol. 57, 155–208, Academic Press.

Wendel, A. M., D. H. Johnson, U. Sharapov et al. 2009. Multistate outbreak of *Escherichia coli* O157:H7 infection associated with consumption of packaged spinach, August–September 2006: The Wisconsin investigation. *Clin. Infect. Dis.* 48: 1079–1086.

Werner, S., K. Boman, I. Einemo et al. 2007. Outbreak of *Salmonella* Stanley in Sweden associated with alfalfa sprouts, July–August 2007. *Euro Surveill.* 12:pii = 3291. http://www.eurosurveillance.org/ViewArticle.aspx?ArticleId = 3291 (accessed December 14, 2009).

Wheeler, C., T. Vogt, G. L. Armstrong et al. 2005. An outbreak of Hepatitis A associated with green onions. *N. Engl. J. Med.* 353: 890–897.

Wismer, W. V. 2009. Consumer eating habits and perspectives of fresh produce quality. In *Postharvest Handling: A Systems Approach*, ed. W. J. Florkowski, R. L. Shewfelt, B. Brueckner, and S. E. Prussia, 23–42, Massachusetts: Academic Press.

4

Managing Food Safety Risks in the Beef and Dairy Industries

4.1 Managing Food Safety Risks in the Beef Industry

4.1.1 Introduction

The major causes of concern and product recalls associated with fresh meat products are typically linked to *Escherichia coli* O157:H7, *Salmonella enteritidis*, *Campylobacter jejuni*, and *Listeria monocytogenes* in ready-to-eat meat and poultry products (Sofos, 2008). Even though most of the foodborne outbreaks were traced to improper food handling practices, Nørrung and Buncic (2008) argued that the original sources of foodborne pathogens that cause most meat-borne bacterial diseases are asymptomatic farm animals that carry and shed pathogens in the feces. In many cases, farmed ruminants carrying zoonotic pathogens in the gastrointestinal tract show no signs of infection (Adam and Brülisauer, 2010; Johnston, 1990). In this chapter, we will focus on the main food safety risks, prevention, and intervention strategies in beef cattle and dairy productions. A case study of *Listeria monocytogenes* in pasteurized milk will be discussed at the end of the chapter.

4.1.2 Main Risks Associated with Beef

The major causes of concern and product recalls associated with beef products is *Escherichia coli* O157:H7. This pathogen is found in animal feces (Hutchinson et al., 2005) and contamination of carcasses and food products by animal feces is likely to be the major method for transmitting foodborne pathogens to consumers (Oliver et al., 2008). The prevalence of *E. coli* O157:H7 in the environment, ability to infect and reinfect cattle as well as a wide host range in wildlife makes complete eradication of *E. coli* O157:H7 an unrealistic goal. Traditional means of controlling infectious agents, such as eradication, testing, and removal of carrier animals are no longer feasible. However, farm management practices—especially the maintenance of quality standards of feed and water—may be the most practical means of reducing infectious agents in cattle on the farm (Hancock et al., 2001). An achievable objective might be to reduce the magnitude or prevalence of *E. coli* O157 in fecal excretion and to break the contamination cycle (Cray et al., 1998; LeJeune and Wetzel, 2007). The contamination cycle in food-producing animals occurs through the ingestion of microbiologically contaminated feeds and water that are contaminated with feces. The use of untreated manure as fertilizer and spread of slurry on grazing fields tend to encourage the wider spread of microbial pathogens. Stresses on animals due to poor management (Nørrung and Buncic, 2008), quantity and quality of animal feed will increase the susceptibility to infections and shedding of foodborne pathogens (Cray et al., 1998; Adam and Brülisauer, 2010). Oliver et al. (2008) suggested that all these environmental and management factors must be considered when identifying farm practices and critical control points on the farm where contamination occurs. An understanding of the possible sources of on-farm infection is important for effective control.

4.1.3 Sources and Transmission Routes of *E. coli* O157:H7 in Beef Cattle

There are numerous reviews that critically discussing the risk factors and the transmission and prevalence of *E. coli* O157:H7 in cattle (Oliver et al., 2008; Ellis-Iversen et al., 2009; Berry and Wells, 2010). The prevalence and potential transmission of *E. coli* through water, feed, hide, and wild/domestic animals are important.

4.1.3.1 Water and Feed

LeJeune et al. (2001) demonstrated that cattle water troughs can be reservoirs of *E. coli* O157:H7 on farms and serve as a source of infection for cattle. This is in agreement with Hancock et al. (1998) who reported that *E. coli* O157:H7 were able to survive in water trough sediments for at least 4 months and appear to multiply especially in warm weather. However, improved water trough hygiene did not reduce the risk of *E. coli* O157:H7 in young cattle (Ellis-Iversen et al., 2008, 2009). Similar to water contamination, the hygiene of animal feed plays a key role in microbial contamination in livestock (Crump et al., 2002) since feed can be a vehicle for transmitting *E. coli* O157:H7 to cattle (Hancock et al., 2001; Davis et al., 2003; Horchner et al., 2006). Fenlon and Wilson (2000) also demonstrated that *E. coli* can multiply in feed. *E. coli* O157:H7 inoculated in laboratory silage (made from rye grass) increased from an initial level of 10^3 cfu/g to 10^7 cfu/g within 13 days. *E. coli* O157:H7 was isolated from the oral cavities of 74.8% of cattle examined, and it is possible that feed may be contaminated by *E. coli* O157:H7 from cattle saliva (Keene and Elder, 2002). Cattle return rumen contents to their mouths to be chewed again and further digested and this may be the most probable source of *E. coli* O157:H7 in the animals' mouths (Tkalcic et al., 2003). Other possible sources include fecal contamination by wildlife (Fischer et al., 2001; Renter et al., 2001), including birds (Shere et al., 1998; Nielsen et al., 2004), rodents (Nielsen et al., 2004), and insects (Ahmad et al., 2007). However, LeJeune et al. (2006) did not find significant correlation between the magnitude of feed contamination and *E. coli* O157:H7 prevalence in cattle. Diez-Gonzalez et al. (1998) argued that a grain diet promoted acid production in the colon, which leads to increased acid resistance of generic *E. coli* strains. The authors demonstrated that hay-fed cattle had a lower concentration of volatile fatty acids in their colons, and the acid shock killed more than 99.99% of the *E. coli*. When diets were supplemented with grain, acids accumulated, colonic pH declined, and this selectively favors *E. coli* resistant to extreme acid shock. There has been much debate since the intervention was first described. Generic coliforms are different from *E. coli* O157:H7. Even though generic *E. coli* from hay-fed cattle are more sensitive to acid shock than those from grain-fed cattle (Diez-Gonzalez et al., 1998; Hovde et al., 1999), however, the acid resistance of *E. coli* O157:H7 was unaffected by diets (Grauke et al., 2002; Hovde et al., 1999). Van Baale et al. (2004) also demonstrated that feeding forage actually increased the shedding of *E. coli* O157:H7 in cattle. Cattle fed forage diets were culture positive for *E. coli* O157:H7 in the feces for longer duration than cattle fed a grain diet (Van Baale et al., 2004). Judging from the various dietary interventions, Callaway et al. (2009) emphasized that dietary manipulations is a powerful method to reduce *E. coli* populations in cattle and further research is essential to clarify the effect of different diets on the bovine gastrointestinal system (Wood et al., 2006).

4.1.3.2 Role of Hide for Pathogen Transmission

Hide cleanliness and prevalence of foodborne pathogens may be associated with the pen feedlot condition; Smith et al. (2001) observed that higher percentages of cattle in muddy feedlot pens shed *E. coli* O157:H7 compared to cattle in normal pen conditions. These researchers reasoned that the muddy feedlot pens may facilitate fecal–oral transmission. Similarly, Cobbold and Desmarchelier (2002) suggested that pen floors and hides were the main source of STEC transmission to dairy calves. Bach et al. (2005) also suggested that feces on pen floors are a more significant source of infection than feed or drinking water. Reid et al. (2002) found that the brisket contains the highest level of bacteria on the hide compared to the rump or flank area. This is in agreement with McEvoy et al. (2000) who demonstrated that the total viable counts were significantly higher at the brisket. This may coincide with the fact that the brisket area is in contact with the floor when cattle are resting. This is also the site where the initial cut is made during the hide removal process, and there is a high probability of transferring pathogens from the hide to carcass (McEvoy et al., 2000). High-level fecal shedding of *E. coli* O157:H7 has also been linked to increased hide contamination (Arthur et al., 2009; Stephens et al., 2009). It is presumed that super-shedding cattle would have a larger impact on the overall contamination of animals due to the increased animal density and confined spaces associated with farm and lairage environments (Arthur et al., 2010). Cattle that excrete higher numbers of *E. coli* O157 compared to other individuals have been referred to as high-level shedders or "super-shedders" (Berry and Wells, 2010; Chase-Topping et al., 2007). High shedding has been defined as counts of *E. coli* O157 that are $\geq 10^3$ (Low et al., 2005) or 10^4 CFU/g of feces (Ogden et al., 2004; Omisakin et al., 2003). Matthews et al. (2006) reported that 20% of the *E. coli* O157:H7 infections are responsible for 80% of the transmission of the organism in Scottish cattle farms. Another study reported similar findings, where 9% of the animals shedding *E. coli* O157:H7 produced over 96% of the total *E. coli* O157:H7 fecal load for the group (Omisakin et al., 2003). Cattle that did not shed *E. coli* O157:H7 during a study conducted by Cobbold et al. (2007) showed that the cattle were five times more likely to be housed in a pen that did not contain a super-shedder. Matthews et al. (2006) suggested that the spread of *E. coli* O157:H7 between cattle could be controlled if one could prevent super-shedding in the 5% of the individuals that are the main source of contamination. Ultimately, significant reductions may be made by targeting the super-shedders (Chase-Topping et al., 2008). Measures that reduce the carriage and shedding of *E. coli* O157:H7 also have the potential to reduce secondary transmission through feed, drinking water, or direct contact and grooming (Gyles, 2007). Chase-Topping et al. (2008) presented a comprehensive discussion of super-shedding and the risk for human infection.

4.1.3.3 Wild and Domestic Animals and Insects

Wildlife fecal contamination can serve as a potential source of infection to livestock (Daniels et al., 2003). Animals demonstrated to carry *E. coli* O157:H7 include, but are not limited to, wild deer (Sargeant et al., 1999; Fischer et al., 2001; Renter et al., 2001), rats (Cizek et al., 1999; Nielsen et al., 2004), and raccoons (Shere et al., 1998). Birds are another singular and important transmission and birds found positive for *E. coli* O157:H7 including pigeons (Shere et al., 1998), gulls (Wallace et al., 1997), and starlings (Nielsen et al., 2004). A study by Scaife et al. (2006) in Norfolk, UK found 20.62% (20/97 samples) of fecal samples collected from wild rabbits were positive for STEC O157. Ahmad et al. (2007) showed that houseflies are capable of transmitting *E. coli* O157:H7 to cattle. Fecal samples from all calves exposed to inoculated flies were positive. The pathogen counts were as high as 1.5×10^5 CFU per fly. This high concentration of *E. coli* O157:H7 suggested that houseflies are not simply mechanical vectors; but the pathogen likely multiplied in the gastrointestinal tract of the houseflies (Alam and Zurek, 2004). Although cattle are considered the main reservoir of *E. coli* O157:H7, it is possible that strains of *E. coli* O157:H7 are introduced into cattle populations through feed (Daniels et al., 2003) and water contaminated with the feces of wild and domestic animals (Wetzel and LeJeune, 2006). In most instances it is impossible to keep wild animals out of a farm but it is important for farms and farm workers to be aware that wild animals can also act as vectors for infection via the fecal–oral route.

4.1.3.4 Veterinary Drug Residues in Beef

The use of veterinary drugs to manipulate the growth of food animal species is strictly prohibited in the European Union (EU; Council Directive 96/23/EC). Some of these drugs are permitted under specific circumstances (for therapeutic purposes) but are only administered under the control of a veterinarian (Van Peteguem and Daeselaire, 2004). Official monitoring of residues in cattle throughout the EU in 2007 found 0.14% noncompliance (out of 58,000 samples from cattle) for the use of illegal growth promoters (GPs); including sex steroids, corticosteroids, and β-agonists (Nebbia et al., 2011). However, this figure may have underestimated the real incidence of GP abuse in meat cattle breeding. For example, in a study in Italy, β-agonists, corticosteroids and sex steroids were found in 11.7%, 17.7%, and 31.9% of the tissues examined from 295 veal calves and 1035 finishing bulls. The lesions were reported to have developed from the illegal administration of the GP drugs (Italian Ministry of Health 2008; cited by Nebbia et al., 2011). Another major concern is the use of drugs which are not subjected to routine controls. Some measures taken by users include combining low doses of compounds belonging to different groups to enhance the synergistic effects of the drugs (e.g., corticosteroids and β-agonists in one cocktail) or by using products known to be hard to detect or analyzed due to considerable metabolism (e.g., stanozolol; Courtheyn et al., 2002). The usage of "cocktails" (mixtures of low amounts of several substances that exert a synergistic effect) for growth promotion and reduce the prospect of analytical detection (Reig and Toldrá, 2008) is dangerous as these substances may cause adverse effects on consumers. The European Food Safety Authority issued an opinion on the effect of hormones residues in meat and provided evidence of association between red meat consumption and hormone-dependent cancers (i.e., breast, endometrial, and ovarian cancer in females, and prostate and testicular cancer in males; EFSA, 2007). A recent drug intoxication due to consumption of lamb and bovine meat containing residues of clenbuterol in Portugal resulted in 50 intoxicated persons with symptoms such as gross tremors of the extremities, tachycardia, nausea, headaches, and dizziness (Barbosa et al., 2005).

The U.S. Food and Drug Administration (USFDA) approved a number of steroid hormone drugs, which include natural estrogen, progesterone, testosterone, and their synthetic versions for use in beef cattle. These drugs increase the animals' growth rate and the leanness of their meat. The FDA only approves the drugs after extensive studies to ensure the food from the treated animals is safe for consumption and that the drugs do not harm the treated animal or the environment (USFDA, 2011). These steroid hormone drugs are formulated as pellets placed under the skin on the back side of the animal's ear. The pellets dissolve slowly under the skin and do not require removal. The ears of the treated animals are discarded at slaughter and are not used for human consumption. The FDA has established the acceptable safe limits for hormones in meat after extensive study and review. All approved implant products have a zero day withdrawal, meaning that the meat is safe for human consumption any time after the animal is treated.

4.1.4 Prevention and Intervention Strategies

A multistate outbreak of *E. coli* O157:H7 associated with beef from Fairbank Farms occurred in the United States from September–November 2009. Twenty-six persons were affected, where 19 were hospitalized and 5 developed hemolytic uremic syndrome (kidney failure). Two deaths were reported (CDC, 2009a). The fact that microorganisms are widely distributed in foods of animal origin remains a significant challenge for meat industries. Multiple foodborne outbreaks of *E. coli* O157:H7 in beef products had occurred since the 1980s. Although most of the outbreaks were traced to improper food handling practices and consumption of undercooked meat, the key question is where does the source of pathogen originate from? The original sources of foodborne pathogens (i.e., *E. coli* O157:H7) are asymptomatic farmed ruminants that carry and shed the pathogens in the feces. Cattle will remain the primary reservoir for *E. coli* and *E. coli* O157:H7. Beef and meat products will be contaminated due to inappropriate handling practices. One may ask, "Why can't we ask the consumer to just cook it? Why eat rare, undercooked beef products?" The palate and behavioral culture are indeed hard to comprehend. But we will discuss this in detail in Chapter 9. In the meantime, the key question here is how can we prevent or reduce the number of foodborne illnesses due to contaminated beef products? On-farm interventions provide a practical and economical program to reduce the prevalence of foodborne pathogens (and diseases) prior to slaughter. This will then reduce the risk of carcass contamination at slaughter and processing facilities.

It must be noted that *E. coli* and *E. coli* O157:H7 are part of the microbiota in cattle and cannot be eliminated. We can, however, reduce the prevalence and magnitude of contamination at the farm level, during transportation and at the abattoir. Some possible routes of contamination and recontamination at the farm level are summarized:

- Super-shedders shedding copious amounts of *E. coli* O157:H7
- Fecal material contaminating feed, water trough, and hide
- Contaminated feces used as slurry spread on grazing pasture
- Domestic and wild animals including vectors such as flies

Farm management practices—especially the maintenance of feed and water—may be the most practical means of reducing infectious agents in cattle (Hancock et al., 2001). Oliver et al. (2008) suggested that all environmental and management factors must be considered when identifying farm practices and critical control points on the farm where contamination occurs. According to Collins and Wall (2004), it is the primary producer who should take all reasonable measures to reduce the entry and prevalence of *E. coli* O157 on his farm. They are responsible for the health of their stock and must adopt a positive approach to animal health on the farm with the objective of eliminating or minimizing exposure of the livestock to zoonotic agents. It is difficult to pinpoint a single practice to a producer or a feedlot operator in controlling *E. coli* O157:H7 (Koohmaraie et al., 2005). Regardless of the challenges, Loneragan and Brashears (2005) reported that the potential of on-farm control exists. It is expected that implementation of on-farm interventions should reduce the prevalence of *E. coli* O157:H7 carried by cattle entering the rest of the food chain.

4.1.4.1 Feed

A number of studies have identified animal feed as a potential source of infection of cattle with STEC O157 (Hancock et al., 2001; Dodd et al., 2003; Van Donkersgoed et al., 2005). In 2004, *Codex Alimentarius* introduced the Code of Practice on Good Animal Feeding to establish a feed safety system for food-producing animals. The objective of the Code is to help ensure the safety of food through adherence to good animal feeding practices at the farm and good manufacturing practices during processing and handling of feed and feed ingredients. It also states that "where appropriate, Hazard Analysis Critical Control Point (HACCP) principles should be followed to control hazards that may occur."

Dietary intervention has been suggested to offer a simple and practical means of reducing the prevalence of *E. coli* O157:H7 in the hindgut (Fox et al., 2007). Berg et al. (2004) showed that cattle fed corn-based diets shed more generic *E. coli* than do cattle fed barley-based finishing diets. The more extensively cereal grains were processed, the more starch was digested in the rumen and this reduced the amount entering the lower digestive tract (Huntington, 1997). Since corn is less digestible in the rumen compared to barley (Huntington, 1997), this provided more undigested starch in the large intestine and resulted in increased fermentation and reduced fecal pH. Corn-fed cattle were found to have lower fecal pH values (5.85) compared to barley-fed cattle (6.51) (Berg et al., 2004; Buchko et al., 2000). Feeding dry-rolled grain diet to cattle reduced the prevalence of *E. coli* O157:H7 by 35% as compared to steam-flaked grain diet. It is possible that dry-rolling allows more substrate to reach the hindgut where it increased fermentation and volatile fatty acid production and made the hindgut inhospitable to the survival of *E. coli* O157:H7 (Fox et al., 2007; Depenbusch et al., 2008).

The fermentation of cereal grains to produce ethanol results in a co-product called *distillers' grains* (DG). The co-product is fed either as wet distillers' grain (WDG) (approximately 30% dry matter) or dried distillers' grains (DDG) (approximately 90% dry matter) (Spiehs et al., 2002). Distillers' grains were shown to increase daily weight gain in finishing cattle due to the condensed nutrients and hence were used in ruminant diets (Ham et al., 1994).

Cattle fed diets including 25% of DDG or 40% of WDG with solubles (WDGS) had a higher prevalence of *E. coli* O157:H7 in their feces. It is possible that (i) feeding dried distillers grains results in decreased starch concentration in the hindgut, which may alter the ecology and favor the growth of *E. coli* O157:H7 or (ii) components of the brewers' grain may stimulate the bacterial growth (Jacob et al., 2008; Wells et al., 2009). Cattle fed 20% or 40% of WDGS were also found to have prolonged survival of inoculated *E. coli* O157:H7 compared to those fed 0% WDGS. The slurries obtained from cattle fed 20% or 40% WDGS had lower concentrations of L-lactate and pH values between 6.0 and 8.0 (Varel et al., 2008). L-lactate has a significant antimicrobial effect on *E. coli* O157:H7 and non-O157 *E. coli* (McWilliam Leitch and Stewart, 2002).

4.1.4.2 Probiotics and Direct-Fed Microbials

Probiotics or direct-fed microbials are preparations of live bacteria fed to a host to elicit beneficial health effects in the host (Schrezenmeir and de Vrese, 2001). Several lactic acid bacteria (LAB), most commonly *Lactobacillus*, *Enterococcus*, and *Streptococcus* have been tested as probiotic agents or competitive exclusion products (CEP) for livestock (Brashears et al., 2003b). Competitive exclusion cultures consist of a mixture of undefined microbes and are usually isolated from the gastrointestinal tract of the animal species that will be treated, while probiotics are well-defined strains that have been cultured separately prior to application (Doyle and Erickson, 2011).

The genus *Lactobacillus* is one of the most commonly used probiotic genera added to a range of feeds (Gaggia et al., 2010). A specific strain, *Lactobacillus acidophilus* NP51, reduced the prevalence of *E. coli* O157:H7 by 49% in animals receiving NP51 compared to controls (Brashears et al., 2003a). Peterson et al. (2007) reported that by administering *L. acidophilus* strain NP51 in feed daily for 2 years, fecal shedding of *E. coli* O157:H7 decreased by 35% in beef cattle. In another study, steers fed a standard steam-flaked corn-based finishing diet containing *L. acidophilus* NP51 showed a reduction of *E. coli* O157:H7 fecal shedding by 57% (Younts-Dahl et al., 2004) while a combination of *L. acidophilus* NP51 and *Propionibacterium freudenreichii* reduced fecal shedding of *E. coli* O157 by 32% compared to the control group (Tabe et al., 2008).

4.1.4.3 Prebiotics

Prebiotics are "nondigestible food ingredients such as fructooligosaccharides (FOS), inulin and galactooligosaccharides (GOS), that beneficially affect the host by selectively stimulating the growth and/or activity of one or a limited number of bacteria in the colon" (Gibson and Roberfroid, 1995; Gaggia et al., 2010). Oligosaccharides are of interest because they are neither hydrolyzed nor absorbed in the upper part of the gastrointestinal tract but stimulate the growth and/or activity of desirable bacteria in the colon (Cummings et al., 2001). The use of prebiotics in cattle has been limited due to the ability of ruminants to degrade most prebiotics, but developments in rumen-protective technologies may allow prebiotics to be used in feedlot and dairy cattle (Callaway et al., 2008b; Doyle and Erickson, 2011). There is a concern that prebiotics may promote satiety and this may decrease feed intake and weight gain in animals (Cani et al., 2005). Ultimately, cattle in feedlots need to be fed energy-dense diets to improve growth and produce a high quality product and any pathogen reduction benefit from a diet should not come at an increased cost for the farm (Berry and Wells, 2010).

Essential oils have been shown to inhibit foodborne pathogens in pure culture (Burt, 2004). The addition of plant phenolic acids such as cinnamic acid, coumaric acid, and ferulic acid to feces increased the death rate of *E. coli* O157:H7 (Wells et al., 2005). Addition of orange peel and pulp to inoculated ruminal fluid reduced *E. coli* O157:H7 from 10^5 to 10^2 CFU/ml. This may be the result of the antimicrobial action of essential oils such as limonene found in the peel (Callaway et al., 2008a). It is still unknown which constituents or mixtures of essential oils are responsible for their antimicrobial activity (Espina et al., 2011). The major chemical component of most citrus oils is limonene with sweet orange containing 68–98% and lemon 45–76% (Svoboda and Greenaway, 2003). Further research is still needed to determine the mechanisms of action and whether the antimicrobial activity can be expressed in the livestock's lower gastrointestinal tract (Callaway et al., 2008a). In addition, candidate plant compounds with antimicrobial activity or grasses used as cattle forages that contain phenolic acids can be used as potential dietary additives or manure treatments (Wells et al. 2005). However, Doyle and Erickson (2011) suggested that the active components of antimicrobial compounds may not be reaching the *E. coli* colonization sites in animals; hence, encapsulation of these ingredients may warrant further investigation. Numerous studies have been carried out on dietary interventions to determine the optimum method of reducing the prevalence of *E. coli* O157:H7 in beef cattle. Since *E. coli* O157:H7 is a normal flora of cattle, it is a daunting task to reduce its presence in the intestine.

4.1.4.4 Husbandry

In the United Kingdom, cattle are graded before slaughter according to a 5-point cleanliness scoring system. This is in accordance with the Meat Hygiene Service's Clean Livestock Policy with lower scores of 1–2 given to clean and dry animals and scores of 4–5 given to filthy and wet animals. Only livestock classed as category 1 and 2 (clean and dry/slightly dirty and dry/damp) are allowed to proceed to slaughter without further interventions (FSA, 2007). This underscores the importance of hide cleanliness, which helps to minimize the transfer of pathogens to the carcass during dressing. Providing sufficient clean and dry bedding will be the most effective means of preventing heavy soiling of the brisket area (Reid et al., 2002). In another study, the housing of cattle in pens surfaced with pond ash (a type

---

BOX 4.1 TEN RISK REDUCTION POINTS FOR FOODBORNE PATHOGEN CONTROL IN A CATTLE FARM

- Provide clean and dry bedding
- Empty and clean water troughs every 2–3 weeks
- Pest control
- Hay or cereal finishing diets (dry diets)
- Source feed from approved suppliers and manufacturers that follow Good Manufacturing Practices
- Allow a down-time period between manure spreading and grazing
- Avoid young stock contact between herds
- Suitable stocking densities
- Adequate building design and layout
- Use of scrapers

---

of by-product from coal combustion) or pens surfaced with soil did not affect fecal shedding of *E. coli* O157:H7 by cattle (Berry et al., 2010). An adequate design and layout of the resting and feeding area and the use of scrapers are also essential hygienic measures. However, Barker et al. (2007) reported that even though automatic scrapers can improve hygiene in barns because of frequent scraping, they can also make cattle dirtier because of the wave of slurry that coats the claws and possibly the lower legs of cattle. The exclusion of wild animals from livestock is beneficial since it is possible that *E. coli* O157:H7 may be introduced into cattle populations through the environment, feed, and water contaminated with wild animals' feces (Daniels et al., 2003; Synge et al., 2003; Wetzel and LeJeune, 2006). Box 4.1 shows the ten risk reduction points for foodborne pathogen control in a cattle farm.

4.2 Managing Risks in the Dairy Industry

4.2.1 Introduction

Milk from cows and other mammals has been an important source of nutrients for humans. Traditionally, individual families or villages tended to one or two animals such as cows, goats, or buffaloes, which provided milk for immediate consumption (Manners and Craven, 2003). Over time, the number of animals increased and production practices evolved to cater to the increasing herd. It is predicted that milk and dairy consumption increase from 78.3 kg/person/year in 2001 to 92 kg/person/year in 2030 and 100 kg/person/year in 2050. Dairy production is highest in India (110 million ton), United States (85 million ton), and China (40 million tons; FAOSTAT, 2010). Meanwhile, the main dairy importers are Italy and Germany (>$1 billion), followed by Belgium and France (>$600 million), and the Netherlands ($450 million). Similarly, the main dairy exporters are Germany (>$1 billion), followed by France (>$ 800 million) and Belgium (>$ 500 million) and the Netherlands and Austria (>$ 400 million; FAOSTAT, 2010).

4.2.1.1 Milkborne Outbreaks and Sources of Contamination

A number of outbreaks had been investigated in which raw milk was implicated as the vehicle of transmission. In 2006, public health officials reported a total of 1270 foodborne outbreaks from 48 states in the United States. Although the dairy commodity accounted for only 3% of single commodity outbreak-related cases (16 outbreaks and 193 cases), 71% of dairy outbreak cases were attributed to unpasteurized (raw) milk (10 outbreaks and 137 cases). Different bacterial pathogens were associated with the outbreaks: *Campylobacter* (six outbreaks), STEC O157 (two outbreaks), *Salmonella* (one outbreak), and *Listeria* (one outbreak), resulting in 11 hospitalizations and one death (CDC, 2009b). In England and Wales, the most common vehicle of infection associated with milkborne outbreaks between 1992 and 2000 was unpasteurized milk (52%); followed by pasteurized milk (37%) with most of the outbreaks (67%) linked to farms. The most common pathogens detected in the outbreaks were *Salmonella* (37%), STEC O157 (33%) and *Campylobacter* (26%) (Gillespie et al., 2003).

Table 4.1 shows a summary of milkborne-related outbreaks. Most of the outbreaks were due to consumption of raw milk (Figure 4.1). There were also cases of contaminated pasteurized milk, which resulted in three deaths and one stillbirth in 2007. Moreover, a large staphylococcal food poisoning outbreak that occurred in Japan in year 2000 was due to consumption of low-fat milk produced from skim milk powder contaminated with *S. aureus* enterotoxin A. Enterotoxins produced by enterotoxigenic strains of *S. aureus* can result in staphylococcal food poisoning in man (Zschöck et al., 2005). It was found that pasteurization had destroyed staphylococci in milk but the enterotoxin had retained enough activity to cause intoxication (Asao et al., 2003).

4.2.1.2 Main Risks Associated with Milk

Cattle can be a reservoir of foodborne pathogens such as *Campylobacter* spp., Shiga toxin-producing *Escherichia coli* (e.g., *E. coli* O157:H7), *Listeria monocytogenes*, *Salmonella* spp. (including multidrug-resistant strains), *Yersinia* spp., *Mycobacterium bovis,* and *Brucella* spp. Nontyphoidal *Salmonella* spp. and *Campylobacter jejuni* are considered important threats because of the high number of illnesses they can cause. *Listeria monocytogenes* and *Escherichia coli* O157:H7 are important pathogens due to the severity of symptoms and number of deaths in infected people. Typically, milk in the mammary gland does not contain bacteria (unless there is intramammary infection or an

Managing Food Safety Risks in the Agri-Food Industries

TABLE 4.1

Outbreak of Infections Associated with Milk

Year	Location (U.S. State or Country)	Type of Pathogen	Cases (Hospitalizations/ Deaths)	Source/Cause of Contamination	Reference
March–August 2011	Pennsylvania	*Yersinia enterocolitica*	16 (7 hospitalized)	Pasteurized milk and ice cream processed in the dairy; Mechanism of contamination unknown and milk and environmental samples from the dairy tested negative	CDC, 2011
April–June 2010	Utah	*Salmonella enteria* serotype Newport	10 (1 hospitalized)	Consumed unpasteurized milk	CDC, 2010
June 2007	The Netherlands	*Campylobacter jejuni*	16	Drank raw milk at dairy farm; *C. jejuni* isolated from bulk milk tank was similar to patients' isolates	Heuvelink et al., 2009
October 2007	Kansas	*Campylobacter jejuni*	68 (2 hospitalized)	Cheese made from unpasteurized milk	CDC, 2009c
October–November 2007	Massachusetts	*Listeria monocytogenes*	5 (5 hospitalized; 3 deaths; 1 stillbirth)	Consumed contaminated pasteurized milk	CDC, 2008a
September 2006	California	*Escherichia coli* O157:H7	6 (2 with hemolytic uremic syndrome, kidney failure)	Contaminated raw milk and raw colostrum	CDC, 2008b

April 2005	The Netherlands	*Campylobacter jejuni*	22	Contaminated raw milk implicated as vehicle of transmission; *C. jejuni* was not isolated from the bulk tank milk but identical strains isolated from fecal samples of dairy cattle	Heuvelink et al., 2009
November 2002–February 2003	Multistate	*Salmonella enterica* serotype Typhimurium	62	Consumed raw milk from dairy (a combination of dairy–restaurant–petting zoo)	Mazurek et al., 2004
December 2001	Wisconsin	*Campylobacter jejuni*	75	Drank unpasteurized milk from local dairy—obtained as milk samples during community events, farm tours, and through cow-leasing program	CDC, 2002
2000	Kansai district, Japan	Staphylococcal enterotoxin A	13,420	Consumed contaminated low-fat milk and powdered skim milk. The thermal process killed the staphylococci in milk but the toxin remained to cause intoxication	Asao et al., 2003

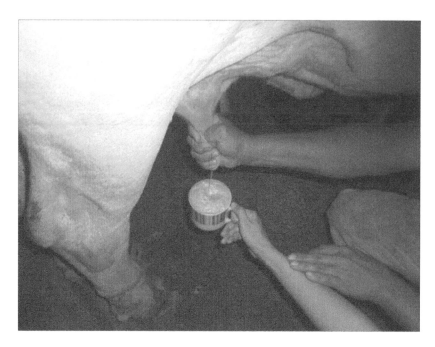

FIGURE 4.1
Raw milk latte anyone? (Photo with permission from M. Kano, 2011.)

animal has a systemic disease). However, as milk is excreted, it can become contaminated with bacteria that colonize the teat. Prevalence of foodborne pathogens in dairy cows and dairy farm environments are shown (Table 4.2). Most of the reported samples were obtained from bulk tank milk. Bulk tank milk analysis is useful to evaluate milk quality and to monitor the udder health status in a herd. It is also less expensive, more convenient, and faster than testing milk samples from individual or groups of cows (Jayarao and Wolfgang, 2003). Bulk tank milk may be able to indicate the level of risk to the public, but Hassan et al. (2000, 2001) revealed that it was not a suitable sample for their studies. The dilution effect from the volumes of milk reduces the chance for detecting a pathogen that exists in low numbers; hence, the researchers obtained samples from the milk filters.

The prevalence of *L. monocytogenes* in bulk tank milk has been reported to range from 1% to almost 19% (Table 4.2). *Listeria monocytogenes* is ubiquitously present in the farm environment. This includes water, grass, silage, decomposing organic matter, soil, and feces (Hassan et al., 2001), and the main source of infection for ruminants are usually spoiled silage (Low and Donachie, 1997; Wiedmann, 2003). *Listeria monocytogenes* can cause mastitis in cows, and it can be shed in milk of asymptomatic cows (Low and Donachie, 1997). The ability of *L. monocytogenes* to survive and proliferate in foods stored at refrigeration temperatures make this organism of particular concern for the dairy food industry, since low initial contamination levels

TABLE 4.2

Prevalence of Foodborne Pathogens in Dairy Cows and Dairy Farm Environments

Foodborne Pathogen	Positive Samples/ Total Number of Samples (%)	Samples	Reference
Campylobacter jejuni	5/248 (2.2)	Bulk tank milk	Jayarao et al., 2006
	8/1720 (0.47)	Raw milk	Steele et al., 1997
	12/131 (9.2)	Bulk tank milk	Jayarao and Henning, 2001
Coxiella burnetii	298/316 (94.3)	Bulk tank milk	Kim et al., 2005
	51/207 (24.6)	Milk obtained from cattle with reproductive disorders	To et al., 1998
Shiga toxin-producing *Escherichia coli*	49/120 (41)	Raw milk	Mhone et al., 2011
	6/248 (2.4)	Bulk tank milk	Jayarao et al., 2006
	2/268 (0.78)	Bulk tank milk	Murinda et al., 2002a
	8/415 (1.93)	Cull dairy cow fecal samples	Murinda et al., 2002a
	312/930 (33.5)	Raw milk	Chye et al., 2004
	22/2144 (12.1)	Milk obtained from dairy cattle with mastitis	Lira et al., 2004
	15/1720 (0.87)	Raw milk	Steele et al., 1997
	5/131 (3.8)	Bulk tank milk	Jayarao and Henning, 2001
Listeria monocytogenes	57/298 (19)	Dairy farm environment (feces, milk, silage, soil, water)	Fox et al., 2009
	3/248 (1.2)	Bulk tank milk	Jayarao et al., 2006
	56/860 (6.5)	Bulk tank milk	Van Kessel et al., 2004
	18/930 (1.9)	Raw milk	Chye et al., 2004
	47/1720 (2.7)	Raw milk	Steele et al., 1997
	6/131 (4.6)	Bulk tank milk	Jayarao and Henning, 2001
	51/404 (12.6)	Bulk tank milk	Hassan et al., 2000
	6/98 (6.1)	Bulk tank milk	Vilar et al., 2007
	9/97 (9.3)	Fecal samples from lactating cows	Vilar et al., 2007
	5/83 (6.0)	Silage samples	Vilar et al., 2007
	3/294 (1.0)	Bulk tank milk	Waak et al., 2002
	23/474 (4.9) in 2000 33/474 (7.0) in 2001	Bulk tank milk	Muraoka et al., 2002

Continued

TABLE 4.2 (*Continued*)

Prevalence of Foodborne Pathogens in Dairy Cows and Dairy Farm Environments

Foodborne Pathogen	Positive Samples/ Total Number of Samples (%)	Samples	Reference
Salmonella spp.	576/2565 (22.5) (S. Newport—41%)	Fecal samples of dairy cattle	Cummings et al., 2009
	15/248 (6)	Bulk tank milk	Jayarao et al., 2006
	22/860 (2.6)	Bulk tank milk	Van Kessel et al., 2004
	13/930 (1.4)	Raw milk	Chye et al., 2004
	3/1720 (0.17)	Raw milk	Steele et al., 1997
	8/131 (6.1)	Bulk tank milk	Jayarao and Henning, 2001
	6/404 (1.5)	Bulk tank milk	Hassan et al., 2000
	6/268 (2.24)	Bulk tank milk	Murinda et al., 2002b
	9/415 (2.17)	Fecal samples	Murinda et al., 2002b
Staphylococcus aureus	91/120 (76)	Raw milk	Mhone et al., 2011
Staphylococcus spp.	14/96 (14.6)	Raw milk sold through unregulated channels	Donkor et al., 2007
Yersinia enterocolitica	3/248 (1.2)	Bulk tank milk	Jayarao et al., 2006
	8/131 (6.1)	Bulk tank milk	Jayarao and Henning 2001
Yersinia spp.	19/96 (19.8)	Raw milk sold through unregulated channels	Donkor et al., 2007

(e.g., <1 cfu/25g) may increase to hazardous levels (Wiedmann, 2003). *Listeria monocytogenes* remains the biggest culprit in raw milk cheeses. The serious consequences of listeriosis such as septicemic form of the illness in elderly and immunocompromised people and abortion in pregnant women or death of their newborn constitute a threat to public health (Latorre et al., 2010).

Salmonella spp. was found in 0.2% to 6% of the raw milk samples (Table 4.2). In Hassan et al. (2000), the researchers isolated *Salmonella enterica* serotype Typhimurium definitive type 104 (DT 104) from one of the positive samples. *Salmonella* Typhimurium DT104 is of major concern to public health due to its multiple antimicrobial resistances (resistant to ampicillin, chloramphenicol, streptomycin, sulfamethoxazole, and tetracycline) (Besser et al., 2000). Gupta et al. (2003) showed that *Salmonella* enterica serotype Newport was resistant to extended-spectrum cephalosporins (antimicrobial drug used to treat salmonellosis). The use of antimicrobial drugs on farms has been associated with the emergence of antimicrobial-resistant salmonella infections (Threlfall et al., 2000; Spika et al., 1987).

Shiga toxin-producing *E. coli* (STEC) was isolated in 0.78–41% of the bulk tank milk samples (Table 4.2). STEC remains a major challenge in dairy farm since ruminants are the main reservoir for *E. coli*. As discussed in the beef section, the fecal–oral cycle is the main source of continuous contamination at the farm level. Also, the presence of super-shedders on the farm will increase the likelihood of *E. coli* O157:H7 infecting other cows. A number of bacteria including *Campylobacter* spp., *Staphylococcus aureus*, and *Yersinia* spp. were also detected in bulk tank milk samples. *Coxiella burnetti* infection in dairy cattle has also been documented and infected cattle can shed enormous numbers of the organisms in milk, birth fluid, and placenta (To et al., 1998; Kim et al., 2005).

4.2.1.3 Veterinary Drugs in Milk

Veterinary drugs are mainly used in the dairy industry to prevent or treat udder infections. However, treatment of lactating cows with veterinary drugs may lead to residues appearing in milk. The presence of veterinary drug residue in milk may result in allergic reactions in sensitized individuals and alter the composition of human intestinal flora. Maximum residue limits (MRLs) have been established by the European Commission (EC, 1996) and Codex (CAC, 2011), while in the United States tolerance levels are fixed in milk (Table 4.3).

TABLE 4.3

Codex and EU Maximum Residue Limits (MRLs) and U.S. Tolerances for 10 Antimicrobial Agents in Bovine Milk

Veterinary Drug Compound	Concentration (µg/l)		
	Codex MRL[a]	EU MRL[b]	U.S. Tolerance[c]
Penicillin	4	4	0[d]
Ampicillin	—	4	10
Amoxicillin	—	4	10
Cephapirin	—	60	20
Ceftiofur	100	100	100
Cloxacillin	—	30	10
Sulfadiazine	—	100	—
Oxytetracycline	100	100	300[e]
Neomycin	1500	1500	150
Erythromycin	—	40	0[d]

[a] CAC (2011)
[b] EC (1996)
[c] Stead et al. (2008)
[d] Indicates tolerances for residues of penicillin and erythromycin in food are established as zero.
[e] Indicates sum of residues of the tetracyclines: chlortetracycline, oxytetracycline, and tetracycline.

4.2.2 Farm Risk Factors Associated with Milk Safety

According to LeJeune and Rajala-Schultz (2009), there are two main risk factors that affect milk safety: the inclusion of organisms in excreted milk (preharvest) and the contamination of milk during collection, processing, distribution, and storage (postharvest). Hence, in order to produce safe and quality milk, farmers should be aware of the risk factors that could lead to contamination of raw milk and how the contamination can be prevented or reduced. Microbial contamination of raw milk may occur through three main sources:

   i. From within the udder (mastitis-associated organisms);

   ii. From environmental organism transfer via dirty udder and teat surfaces;

   iii. From improperly cleaned and sanitized milking equipment (Elmoslemany et al., 2009a).

4.2.2.1 Dairy Farm Hygiene Practices

Listeria, *Salmonella*, and pathogenic *E. coli* are frequently isolated from dairy cattle and farm environment such as water, feed, manure, and bird droppings (Van Kessel et al., 2004). The association between wild birds and salmonellosis infections had been reported. For example, there was a previous outbreak of salmonellosis in a dairy herd due to contamination of hay by wild birds (Glickman et al., 1981). Evans and Davies (1996) conducted a case-control study in the United Kingdom and found that stored feed contaminated by wild birds was a risk factor for *S. Typhimurium* DT104. It was noted that the risk of contamination increased in farms with low-quality silage (pH $\geq$ 4.5), poor milking practices (Vilar et al., 2007), and poor barn hygiene (Sanaa et al., 1993).

4.2.2.2 Animal Udder Hygiene

Elmoslemany et al. (2009a, b) found strong association between poor teat-end cleanliness and failure in washing milking machine appropriately with an increased level of bacterial count. Dirty teats and udders are considered one of the main sources of pathogenic bacteria in milk. Between milking, the teats and udder can become soiled with manure and bedding materials (Chambers, 2002).

4.2.2.3 Fecal–Oral Contamination Route

The fecal–oral contamination route has been suggested by Oliver et al. (2005) in which foodborne pathogens in the dairy farm environment contaminates the feed followed by ingestion and amplification of the pathogen in the cow.

Subsequently, more pathogen is disseminated in the farm environment via feces and through spreading of cattle manure onto croplands.

4.2.2.4 Workers and Farm Visitors

Farms with part-time workers (working up to 25 hours per week) were more likely to be found positive with *Salmonella*. People and animals are known to transfer organisms between farms, and it is possible that part-time farm workers are more likely to be employed on more than one farm, and therefore pose greater risk, than full-time workers. Improvements in the biosecurity measures used by farm staff and visitors before entering and leaving farms should be considered. Changing into clean footwear and overalls and the use of foot-dips and hand washing are common practices on pig and poultry farms, but rarely used on dairy farms (Davison et al., 2006). Farm staff and family members who are in contact with foodborne pathogens-infected livestock and milk are at risk of contracting the disease and carrying the infection within the farm and to other farms. Biosecurity-related factors such as lack of clean visitor parking area, introduction of new adult cattle, and silage storage were associated with an increase prevalence of *Salmonella* in dairy farm herds in England and Wales (Davison et al., 2006).

4.2.3 Interventions to Reduce Contamination at the Farm Level

Animal health, improved milk hygiene, and pasteurization are keys to ensuring milk safety (LeJeune and Rajala-Schultz, 2009). Mastitis control programs have been widely used for at least 60 years to ensure hygienic harvest of milk. Farmers were able to reduce contagious mastitis by adopting the five basic principles of mastitis control: postmilking teat disinfection, universal dry cow antibiotic therapy, appropriate treatment of clinical cases, culling of chronically infected cows, and regular milking machine maintenance (Ruegg, 2003). Even though the five-point mastitis control plans have been widely adopted, there are a number of practical on-farm management interventions that can be adopted to enhance dairy food safety. Raw milk that is not pasteurized and used to manufacture cheese and other dairy products may constitute risk to human health. In order to reduce the contamination at the farm, good farm practices (e.g., animal and waste management), water treatment, and good hygienic conditions during milking and mastitis control are carried out (Kousta et al., 2010).

4.2.3.1 Animal Health and Udder Hygiene

Not even the best cleaning method removes all bacteria but Good Udder Practices is of great importance to reduce pathogen load and consequently milk contamination. Cleanliness of the udder and teats can be influenced by several factors such as (1) transition from summer grazing to winter housing where housed cows are dirtier than grazing cows (Ellis et al., 2007), (2) fecal consistency (where dirtier cows have higher fluid fecal consistency), (3) frequency of bedding change and quality of bedding (Ward et al., 2002), and (4) stage of lactation (Reneau et al., 2005; Ward et al., 2002). Cows during the early lactation stage are usually heavily contaminated with feces.

Udder hygiene scoring systems can be used on farms to evaluate cleanliness and reduce teat contamination. Udder hygiene scoring charts have been developed by Cook (2002), Schreiner and Ruegg (2003), and Reneau et al. (2005). Figure 4.2 shows a hygiene scoring card which documents the degree of manure contamination on a 1–4 scale for each of three zones: the udder, the lower leg, and the upper leg and flank (Cook, 2002). The source of contamination is often the housing area of the cows. The amount of manure present in the back of cow stalls, the dirtiness of the milking parlor, and the thickness of bedding can be routinely checked. Fox et al. (2009) revealed that dairy farms with the best hygiene scores had the lowest percentage of positive *L. monocytogenes* samples. Improvements in hygiene scores for facilities and udders can be used as a key indicator of on-farm risk reduction programs (Ruegg, 2003).

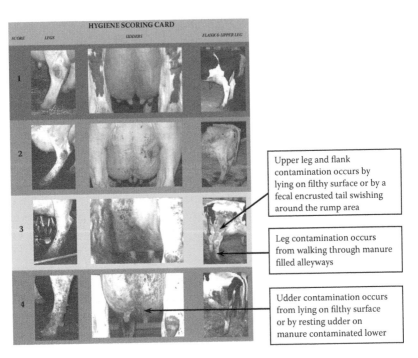

FIGURE 4.2
Udder hygiene scoring chart on a scale of 1–4, with score 1 and 2 showing no or little contamination and scores 3 and 4 showing distinct contamination. (Adapted from: Cook, N. B. 2002. The influence of barn design on dairy cow hygiene, lameness, and udder health. *Proceedings of the 35th Annual Convention of America Association of Bovine Practice*, Madison, Wisconsin, American Association of Bovine Practice. With permission.)

4.2.3.2 Premilking Sanitation

Premilking hygiene and sanitation procedure will reduce the microbial contamination immediately before the attachment of milking units and decrease udder infections (Ruegg and Dohoo, 1997). There were a number of premilking udder preparation methods. To avoid transferring pathogens from cow to cow, each towel should only be used for one cow, and washable towels should be washed at 90°C before use. Cleaning teats and udders with a moist towel is also better than cleaning with a dry paper towel (Galton et al., 1986; Pankey, 1989). Elmoslemany et al. (2010) also found that predipping followed by drying the teats with single-use towel was associated with the lowest bacterial counts. The drying step is an important regime because if teats are washed without drying, the water on the udder and teat surfaces can enter the teat cup liners and increase bacterial contamination of milk (Galton et al., 1982).

Using dip or spray with a soap solution before drying with a paper towel also reduces pathogen load. Iodophor, sodium hypochlorite, and dodecyl benzene sulfonic acid teat disinfectant dips significantly reduced total bacteria in milk with no significant differences among disinfectants (Galton et al., 1986). Predipping has been shown to reduce the risk of *L. monocytogenes* in milk filters by almost fourfold (Hassan et al., 2001). However, premilking disinfectant teat dipping with iodophor can result in higher iodine residue in milk (Galton et al., 1984). Foremilking is important since the first milk from a quarter is lower in quality and high in bacteria—so foremilking removes milk that has a higher concentration of bacteria from contaminating bulk tank milk (Hassan et al., 2001).

4.2.3.3 Hygiene of the Milk Machine

The formation of biofilm in milking equipment may be a possible source of pathogenic reservoir in bulk tank milk (Latorre et al., 2009). Latorre et al. (2010) had detected the presence of a bacterial biofilm on the bottom covers of milk meters. *Listeria monocytogenes* has the potential to form biofilms on materials such as stainless steel (Beresford et al., 2001; Norwood and Gilmour 1999), rubber, and plastic (Beresford et al., 2001). These materials are frequently found in milk handling equipment. The ability of *L. monocytogenes* to form biofilms (Harvey et al., 2007) may contribute to its persistence in food processing plants (Thimothe et al., 2004).

Proper cleaning and sanitation of the milking system is crucial to avoid the presence of chemical and biological residues and to produce safe and high quality milk. It involves a combination of thermal, chemical, and physical processes. A deficiency in any of these parameters could result in soil build-up, which provides nutrients for growth and multiplication of bacteria between milking times (Reinemann et al., 2003). The quality of water used in cleaning and sanitizing solutions is important because it affects the cleaning process. It is advisable to use drinking water quality to prevent possible sources of contamination (Vilar et al., 2012). In fact, Elmoslemany et al. (2010) found that farms without a water purification system were more likely to have elevated bacterial counts than farms with it. Hard water can reduce the effectiveness of cleaning chemicals and may lead to formation of films or deposits on the milking system (Cords et al., 2001).

The cleaning process usually starts with a prerinse using water to remove residual milk and other easily soluble deposits. This is followed by alkaline detergents, which are used to remove organic soils such as milk fat and proteins. The cleaning effectiveness of detergents is usually improved as temperature increases and water hardness decreases. The working temperature range for alkaline detergents is between 43°C and 77°C. Subsequent acid rinse is conducted to remove inorganic (mineral) deposits. The disinfection process is applied using chlorine-based disinfectants at concentrations of 100 to 200 ppm of available chlorine. The most widely used chlorine compounds are hypochlorites of calcium (CaOCl) and sodium (NaOCl; Reinemann et al., 2003).

Defects in the milking machine and faulty milking management will also influence udder health. Vacuum level in the milking system can vary through air admission in the system during teat cup attachment. Undesirable vacuum losses and subsequent milking unit fall-offs could affect udder health and cleanliness (Tan et al., 1993).

4.2.3.4 Bedding Materials

The hygiene of milking environment and dairy farm is related to udder health and hygiene. Type of bedding material, provision of fresh bedding material, and removal of wet spots are important to maintaining hygiene and reducing pathogens in the environment (Dohmen et al., 2010). Dairy cattle can spend between 8 to 16 hours of their time lying down per day (Manninen et al., 2002; Tucker et al., 2003), and teats are exposed to environmental pathogens. The microbial load of lying area surfaces is directly related to the bacterial count of the teats; hence the material used as bedding and flooring plays a key role in maintaining hygiene (De Palo et al., 2006). Counts of bacteria in inorganic bedding are usually lower than those in organic bedding, depending on the bacterial strain and type of material (Fairchild et al., 1982).

Moisture, pH, stall cleanliness, and availability of nutrients play a major role in determining bedding quality. Ward et al. (2002) conducted a study on four farms and found that much of the straw stored for bedding were too wet (>15% moisture content). Moreover, the interior temperature of most of the straw beds reached between 15 °C to 45 °C within about a week after the straw bed was changed. The pH of the top layers of straw was between 8.5–9.5, and adding lime daily to the surface failed to reduce the bacterial count. The emphasis here is that temperature, moisture content, and pH values may be conducive for the multiplication of pathogens. Other factors to take into consideration are quantity of straw used and the stocking density. For example, Hughes (2001) suggested 2.5 to 3 tons per high-yielding cow, 2 tons for mid and late lactation cows and 1.5 tons for dry cows. Meanwhile, a stocking density of 6.5 m² of bedded area for each high-yielding cow, 5.6 m² for each cow in mid or late lactation, and 4.6 m² for each dry cow. Sand has been increasingly used as bedding material in North American dairy farms because it is economical, improves udder health, and has advantages for leg health (Bewley et al., 2001; Espejo et al., 2006; Norring et al., 2008).

4.2.3.5 Perform Coliform Counts

Coliform counts should be less than 100 cfu/ml for milk intended to be pasteurized before consumption and less than 10 cfu/ml for raw milk used for direct consumption. Coliforms can incubate in residual films left on milk contact surfaces such as milking pipelines or equipment. Coliform counts greater than 1000 cfu/ml indicates the possibility of incubation and the cleaning procedures should be investigated (Ruegg, 2003).

4.2.3.6 Pasteurization

Pasteurization is the process of heating milk for a predetermined time at a predetermined temperature. The aim of pasteurization is to destroy microorganisms, to improve safety, and the shelf life of milk. The current legal standard for pasteurization of milk in the UK is heat treatment involving High Temperature Short Time (HTST) treatment (at least 161°F [72°C] for 15 s or any equivalent combination) or a pasteurization process using different time and temperature combinations to obtain an equivalent effect (Dairy Products [Hygiene] Regulations, 1995). In the United States, batch or vat pasteurization is carried out at 145°F (63°C) for 30 min or HTST at 161°F (72°C) for 15 s (see Table 4.4; USFDA, 2007). The USFDA (2007) also stated that the term *pasteurization* and similar terms shall mean the process of heating every particle of milk or milk product in properly designed and operated equipment to a specific temperature and time. The following temperatures and its specified corresponding time apply.

Even though pasteurization is an effective method to eliminate foodborne pathogens, unpasteurized milk may be used to manufacture some specialty cheeses. This is because specialty cheeses preserve the use of raw milk as a constituent of the artisan cheese-making process (Murphy et al., 2004), and raw milk remains popular for cheese making because of enhanced organoleptic properties and flavor differences (Baylis, 2009). The consumption of cheese made from unpasteurized milk is also consumed by a segment of the population (Oliver et al., 2005). Unpasteurized (raw) milk cheeses purchased in the United States are produced by a process approved by the FDA (i.e., 60-days ripening) to mitigate harmful effects of bacteria (Colonna et al., 2011). In Europe, cheese manufacture is controlled by the Food Hygiene

TABLE 4.4

Pasteurization Temperature and Its Specified
Corresponding Time

Temperature	Time
63°C (145°F)	30 min
72°C (161°F)	15 s
89°C (191°F)	1.0 s
90°C (194°F)	0.5 s
94°C (201°F)	0.1 s
96°C (204°F)	0.05 s
100°C (212°F)	0.01 s

Source: USFDA. 2007. Grade "A" pasteurized milk
ordinance (2007 revision) section 1 through
section 7. http://www.fda.gov/Food/Food
Safety/Product-SpecificInformation/MilkSafety
/NationalConferenceonInterstateMilkShipments
NCIMSModelDocuments/PasteurizedMilk
Ordinance2007/ucm063876.htm (accessed
January 27, 2012).

Regulations (EC 2004a, 2004b). Faulty pasteurization and recontamination of
pasteurized milk may pose a threat to consumers, too. In addition to using
raw milk for artisanal cheese making, there is a trend where dairy farm
families and workers tend to consumer raw milk. In the United States consumption of raw bulk tank milk (BTM) is a common practice among farm
families. Preference for raw milk consumption was more common among
dairy farm families and employees since it is the cheaper option, and the
farmers generally believe their milk to be risk-free on the basis of routine
test results and to be of better quality than pasteurized milk (Shiferaw et
al., 2000; Hegarty et al., 2002). Meanwhile, among the nonfarming community, a growing number of consumers advocated that raw milk is a healthier
option and preferred to choose raw milk over pasteurized milk (Bren, 2004).
Raw milk may harbor a number of pathogens, such as *Campylobacter* spp.,
Shiga toxin-producing *Escherichia coli* (e.g. *E. coli* O157:H7), *Listeria monocytogenes*, *Salmonella* spp. (including multidrug-resistant strains), *Yersinia* spp.,
Mycobacterium bovis, and *Brucella* spp. (CDC, 2002). The foodborne pathogens
in raw milk originate from the farm environment and direct excretion from
animals infected udder. The pathogens may then contaminate dairy plants
and subsequently other dairy products. Twenty-nine states allow the sale of
raw milk, for example, through direct purchase, cow-sharing, or leasing programs, and as pet food (Oliver et al., 2009). Let's take the case of Wisconsin,
where unpasteurized milk cannot be sold legally to consumers. However,
dairies can distribute unpasteurized milk through a cow-leasing program in
which farmers keep and milk cows owned by individuals. Customers will
pay an initial fee to lease part of a cow. Farm operators will milk the cows

and store the milk from all leased cows together in a bulk tank. The customers can then collect the milk from the farm or have it delivered to their homes (CDC, 2002). In California, the sale of raw milk is legal, thus making this state the largest producer of *certified raw milk* in the United States (Headrick et al., 1997, 1998). Certified raw milk is unpasteurized milk with a total bacterial count below a specified standard but does not guarantee that the milk is free of bacterial pathogens (Jayarao et al., 2006). As a result of milkborne outbreaks (Table 5.1), California enacted legislation in 2008, setting a limit of 10 coliforms/ml for raw milk sold to consumers (CDC, 2008b). In addition to consumption of raw milk, the faulty pasteurization process and milk that were recontaminated after heat treatment may cause food poisoning outbreaks, too. As long as legislation allows the sale and distribution of untreated milk, the risk of milkborne outbreaks remains.

4.2.3.7 Training

In countries where hand milking is the norm, it is important to train farmers and staff working at milk collection centers regarding milk hygiene. The cows' teats and the milkers' hands should be washed carefully before milking starts and containers used for storing and transporting milk should be cleaned each time before and after using (Millogo et al., 2010). There are a number of advocates who argued the benefits of drinking untreated milk and the damaging effects of heat treatment on flavor and nutrients (Pickard [accessed 2012]; Spiegel, 2008).

In addition, the public should continuously be made aware of the risks of consuming unpasteurized milk or products made from raw milk. Since more public and educational visits are made to the farms today, visitors (especially children) should be informed of the risks of consuming raw milk.

4.2.4 Case Study of *Listeria monocytogenes* Outbreak in Pasteurized Milk

The following is a case study adapted from the U.S. Center for Disease Prevention and Control (CDC, 2008a). Between June to November 2007, *Listeria monocytogenes* with indistinguishable PFGE patterns were isolated from five patients. Three of the patients died and one woman experienced stillbirth at 37 weeks' gestation. Another female delivered a healthy but premature infant. Environmental investigations traced the strain back to Dairy A in Massachusetts. Unopened flavored and unflavored milk samples and environmental swab samples were collected from the dairy. One environmental swab from a floor drain in the finished goods area and eight milk samples were tested positive for *L. monocytogenes,* which matched the outbreak strain by PFGE. When the authorities reviewed the dairy's records, the evidence showed that the plant's equipment met the federal standards for time, temperature, and flow for effective pasteurization. However, the

**BOX 4.2   TEN RISK REDUCTION POINTS IN MANAGING
FOODBORNE PATHOGEN IN DAIRY FARMS**

- Provision of clean bedding and dairy farm hygiene (reduce fecal-oral route contamination cycle)
- Clean, pre-dip teats with disinfectant and dry teats with single-use towel before milking
- Pest, wild and domestic animal control
- Use udder hygiene scoring system to evaluate cleanliness and reduce teat contamination
- Cleaning and disinfection of milking equipment. Ensure that the water supply used for cleaning and sanitizing is of drinking quality water.
- Feed stored according to Good Storage Practices and good quality silage
- Pasteurization at 145°F (63°C) for 30 min or HTST at 161°F (72°C) for 15 seconds
- Biosecurity measures of farm to reduce potential transfer of pathogens between farms – especially if part-time workers are employed in more than one farm
- Training of farmers to improve dairy farm hygiene practices
- Raise awareness among the public visiting dairy farms on the potential risks of consuming raw milk

facility did not have an environmental monitoring program for *L. monocytogenes*. This is not required by law, but is usually implemented as best practice by large food processing companies. As a result, in February 2008, dairy A closed its milk processing facility due to the high cost of implementing mitigation strategies.

In this outbreak, it was most likely that physical design of the facility, product flow, and maintenance procedures contributed to the contamination of finished produce. Even though records revealed that the pasteurization process was adequate, it still remains unclear how the pasteurized milk was contaminated after the pasteurization. However, based on the Latorre et al. (2009) study, *L. monocytogenes* may be able to form a biofilm on a stainless steel surface, and the protection provided by the film may enable cross contamination to the treated milk postpasteurization. This may not be a direct on-farm contamination, but it is possible that *L. monocytogenes* may have originated from the dairy cows and had been established within the processing facility. Box 4.2 shows the 10 risk reduction points for foodborne pathogen control in a dairy farm.

References

Adam, K. and F. Brülisauer. 2010. The application of food safety interventions in primary production of beef and lamb: A review. *Int. J. Food Microbiol.* 141: S43–S52.

Ahmad, A., T. G. Nagaraja, and L. Zurek. 2007. Transmission of *Escherichia coli* O157:H7 to cattle by house flies. *Prev. Vet. Med.* 80: 74–81.

Alam, M. J. and L. Zurek. 2004. Association of *Escherichia coli* O157:H7 with houseflies on a cattle farm. *Appl. Environ. Microbiol.* 70: 7578–7580.

Arthur, T. M., D. M. Brichta-Harhay, J. M. Bosilevac et al. 2010. Super shedding of *Escherichia coli* O157:H7 by cattle and the impact on beef carcass contamination. *Meat Sci.* 86: 32–37.

Arthur, T. M., J. E. Keen, J. M. Bosilevac et al. 2009. Longitudinal study of *Escherichia coli* O157:H7 in a beef cattle feedlot and role of high-level shedders in hide contamination. *Appl. Environ. Microbiol.* 75: 6515–6523.

Asao, T., Y. Kumeda, T. Kawai et al. 2003. An extensive outbreak of staphylococcal food poisoning due to low-fat milk in Japan: Estimation of enterotoxin A in the incriminated milk and powdered skim milk. *Epidemiol. Infect.* 130: 33–40.

Bach, S. J., L. J. Selinger, K. Stanford, and T. A. McAllister. 2005. Effect of supplementing corn- or barley-based feedlot diets with canola oil on faecal shedding of *Escherichia coli* O157:H7 by steers. *J. Appl. Microbiol.* 98: 464–475.

Barbosa, J., C. Cruz, J. Martins et al. 2005. Food poisoning by clenbuterol in Portugal. *Food Addit. Contam.* 22: 563–566.

Barker, Z. E., J. R. Amory, J. L. Wright, R. W. Blowey, and L. E. Green. 2007. Management factors associated with impaired locomotion in dairy cows in England and Wales. *J. Dairy Sci.* 90: 3270–3277.

Baylis, C. L. 2009. Raw milk and raw milk cheeses as vehicles for infection by verocytotoxin-producing *Escherichia coli*. *Int. J. Dairy Technol.* 62: 293–307.

Beresford, M. R., P. W. Andrew, and G. Shama. 2001. *Listeria monocytogenes* adheres to many materials found in food-processing environments. *J. Appl. Microbiol.* 90: 1000–1005.

Berg, J., T. McAllister, S. Bach, R. Stilborn, D. Hancock, and J. LeJeune. 2004. *Escherichia coli* O157:H7 excretion by commercial feedlot cattle fed either barley- or corn-based finishing diets. *J. Food Protect.* 67: 666–671.

Berry, E. D. and J. E. Wells. 2010. *Escherichia coli* O157:H7: Recent advances in research on occurrence, transmission, and control in cattle and the production environment. *Adv. Food Nutr. Res.* 60: 67–117.

Berry, E. D., J. E. Wells, T. M. Arthur, B. L. et al. 2010. Soil versus pond ash surfacing of feedlot pens: Occurrence of *Escherichia coli* O157:H7 in cattle and persistence in manure. *J. Food Protect.* 73: 1269–1277.

Besser, T. E., M. Goldoft, L. C. Pritchett et al. 2000. Multiresistant *Salmonella* Typhimurium DT104 infections of humans and domestic animals in the Pacific Northwest of the United States. *Epidemiol. Infect.* 124: 193–2000.

Bewley, J., R. W. Palmer, and D. B. Jackson-Smith. 2001. A comparison of free-stall barns used by modernized Wisconsin dairies. *J. Dairy Sci.* 84: 528–541.

Brashears, M. M., M. L. Galyean, G. H. Loneragan, J. E. Mann, and K. Killinger-Mann. 2003a. Prevalence of *Escherichia coli* O157:H7 and performance by beef feedlot cattle given *Lactobacillus* direct-fed microbials. *J. Food Protect.* 66: 748–754.

Brashears, M. M., D. Jaroni, and J. Trimble. 2003b. Isolation, selection, and characterization of lactic acid bacteria for a competitive exclusion product to reduce shedding of *Escherichia coli* O157:H7 in cattle. *J. Food Protect.* 66: 355–363.

Bren, L. 2004. Got milk? Make sure it's pasteurized. *FDA Consumer* 38: 29–31.

Buchko, S. J., R. A. Holley, W. O. Olson, V. P. J. Gannon, and D. M. Veira. 2000. The effect of different grain diets on fecal shedding of *Escherichia coli* O157:H7 by steers. *J. Food Protect.* 63: 1467–1474.

Burt, S. 2004. Essential oils: their antibacterial properties and potential applications in foods—A review. *Int. J. Food Microbiol.* 94: 223–253.

CAC. 2011. Codex Alimentarius Commission. Maximum residue limits for veterinary drugs in foods. Updated as at the 34th Session of the Codex Alimentarius Commission (July 2007). CAC/MRL 2-2011, pp 1–36. http://www.codexalimentarius.net/vetdrugs/data/MRL2_e_2011.pdf (accessed January 15, 2012).

Callaway, T. R., J. A. Carroll, J. D. Arthington et al. 2008a. Citrus products decrease growth of *E. coli* O157:H7 and *Salmonella* Typhimurium in pure culture and in fermentation with mixed ruminal microorganisms *in vitro*. *Foodborne Pathog. Dis.* 5: 621–627.

Callaway, T. R., T. S. Edrington, R. C. Anderson et al. 2008b. Probiotics, prebiotics and competitive exclusion for prophylaxis against bacterial disease. *Animal Health Res. Rev.* 9: 217–225.

Callaway, T. R., M. A. Carr, T. S. Edrington, R. C. Anderson, and D. J. Nisbet. 2009. Diet, *Escherichia coli* O157:H7, and cattle: A review after 10 years. *Curr. Issues Mol. Biol.* 11: 67–80.

Cani, P. D., A. M. Neyrinck, N. Maton, and N. M. Delzenne. 2005. Oligofructose promotes satiety in rats fed a high-fat diet: Involvement of glucagon-like peptide-1. *Obes. Res.* 13: 1000–1007.

CDC. 2002. Outbreak of *Campylobacter jejuni* infections associated with drinking unpasteurized milk procured through a cow-leasing program—Wisconsin, 2001. *MMWR Morb. Mortal. Wkly. Rep.* 51: 548–549.

CDC. 2008a. Outbreak of *Listeria monocytogenes* infections associated with pasteurized milk from a local dairy—Massachusetts, 2007. *MMWR Morb. Mortal. Wkly. Rep.* 57: 1097–1100.

CDC. 2008b. *Escherichia coli* O157:H7 infections in children associated with raw milk and raw colostrum from cows—California, 2006. *MMWR Morb. Mortal. Wkly. Rep.* 57: 625–628.

CDC. 2009a. Multistate Outbreak of *E. coli* O157:H7 Infections Associated with Beef from Fairbank Farms. http://www.cdc.gov/ecoli/2009/ (accessed December 20, 2011).

CDC. 2009b. Surveillance for foodborne disease outbreaks—United States, 2006. *MMWR Morb. Mortal. Wkly. Rep.* 58: 609–615.

CDC. 2009c. *Campylobacter jejuni* infection associated with unpasteurized milk and cheese—Kansas, 2007. *MMWR Morb. Mortal. Wkly. Rep.* 57: 1377–1379.

CDC. 2010. Notes from the field: *Salmonella* Newport infections associated with consumption of unpasteurized milk—Utah, April–June 2010. *MMWR Morb. Mortal. Wkly. Rep.* 59: 817–818.

CDC. 2011. Notes from the field: *Yersinia enterocolitica* infections associated with pasteurized milk—Southern Pennsylvania, March–August 2011. *MMWR Morb. Mortal. Wkly. Rep.* 60: 1428.

Chambers, J. V. 2002. The microbiology of raw milk. In *Dairy Microbiology Handbook*, ed. R.K Robinson, 39–90. 3rd ed. New York: John Wiley & Sons.

Chase-Topping, M. E., I. J. McKendrick, M. C. Pearce et al. 2007. Risk factors for the presence of high-level shedders of *Escherichia coli* O157 on Scottish farms. *J. Clin. Microbiol.* 45: 1594–1603.

Chase-Topping, M. E., D. Gally, C. Low, L. Matthews, and M. Woolhouse. 2008. Super-shedding and the link between human infection and livestock carriage of *Escherichia coli* O157. *Nat. Rev. Microbiol.* 6: 904–912.

Chye, F. Y., A. Abdullah, and M. K. Ayob. 2004. Bacteriological quality and safety of raw milk in Malaysia. *Food Microbiol.* 21: 535–541.

Cizek, A., P. Alexa, I. Literak, J. Hamrik, P. Novak, and J. Smola. 1999. Shiga toxin-producing *Escherichia coli* O157 in feedlot cattle and Norwegian rats from a large-scale farm. *Lett. Appl. Microbiol.* 28: 435–439.

Cobbold, R. and P. Desmarchelier. 2002. Horizontal transmission of shiga toxin-producing *Escherichia coli* within groups of dairy calves. *Appl. Environ. Microbiol.* 68: 4148–4152.

Cobbold, R., D. D. Hancock, D. H. Rice et al. 2007. Rectoanal junction colonization of feedlot cattle by *Escherichia coli* O157:H7 and its association with supershedders and excretion dynamics. *Appl. Environ. Microbiol.* 73: 1563–1568.

Collins, J. D. and P. G. Wall. 2004. Food safety and animal production systems: controlling zoonoses at farm level. *Sci. Tech. Rev.* 23: 685–700.

Colonna, A., C. Durham, and L. Meunier-Goddik. 2011. Factors affecting consumers' preferences for and purchasing decisions regarding pasteurized and raw milk specialty cheeses. *J. Dairy Sci.* 94: 5217–5226.

Cook, N. B. 2002. The influence of barn design on dairy cow hygiene, lameness, and udder health. *Proceedings of the 35th Annual Convention of America Association of Bovine Practice.* Madison, WI, American Association of Bovine Practice.

Cords, B. R., G. R. Dychdala, and F. L. Richter. 2001. Cleaning and sanitizing milk production and processing. In *Applied Dairy Microbiology*, ed. J. Steele and E. Marth, 547–585. New York: Marcel Dekker.

Council Directive 96/23/EC (1996) Council Directive 96/23/EC of 29 April 1996 on measures to monitor certain substances and residues thereof in live animals and animal products and repealing Directives 85/358/EEC and 86/469/EEC and Decisions 89/187/EEC and 91/664/EEC. *Off. J. Eur. Comm.* L125/10. http://ec.europa.eu/food/food/chemicalsafety/residues/council_directive_96_23ec.pdf (accessed March 12,2012).

Courtheyn, D., B. Le Bizec, G. Brambilla et al. 2002. Recent developments in the use and abuse of growth promoters. *Analytica Chimica Acta* 473: 71–82.

Cray, Jr. W. C., T. A. Casey, B. T. Bosworth, and M. A. Rasmussen. 1998. Effect of dietary stress on fecal shedding of *Escherichia coli* O157:H7 in calves. *Appl. Environ. Microbiol.* 64: 1975–1979.

Crump, J. A., P. M. Griffin, and F. J. Angulo. 2002. Bacterial contamination of animal feed and its relationship to human foodborne illness. *Clin. Infect. Dis.* 35: 859–865.

Cummings, J. H., G. T. MacFarlane, and H. N. Englyst. 2001. Prebiotic digestion and fermentation. *Am. J. Clin. Nutr.* 73: 415–420.

Cummings, K. J., L. D. Warnick, and K. A. Alexander et al. 2009. The incidence of salmonellosis among dairy herds in the northeastern United States. *J. Dairy Sci.* 92: 3766–3774.

Dairy Products (Hygiene) Regulations. 1995. Schedule 5. Requirements for milk used for the manufacture of milk-based products. Part III Pasteurized milk. http://www.legislation.gov.uk/uksi/1995/1086/contents/made (accessed January 27, 2012).

Daniels, M. J., M. R. Hutchings, and A. Greig. 2003. The risk of disease transmission to livestock posed by contamination of farm stored feed by wildlife excreta. *Epidemiol. Infect.* 130: 561–568.

Davis, M. A., D. D. Hancock, D. H. Rice et al. 2003. Feedstuffs as a vehicle of cattle exposure to *Escherichia coli* O157:H7 and *Salmonella enterica*. *Vet. Microbiol.* 95: 199–210.

Davison, H. C., A. R. Sayers, R. P. Smith et al. 2006. Risk factors associated with the *Salmonella* status of dairy farms in England and Wales. *Vet. Rec.* 159: 871–880.

Depenbusch, B. E., T. G. Nagaraja, J. M. Sargeant, J. S. Drouillard, E. R. Loe, and M. E. Corrigan. 2008. Influence of processed grains on fecal pH, starch concentration, and shedding of *Escherichia coli* O157 in feedlot cattle. *J. Anim. Sci.* 86: 632–639.

De Palo, P., A. Tateo, F. Zezza, M. Corrente, and P. Centoducati. 2006. Influence of free-stall flooring on comfort and hygiene of dairy cows during warm climatic conditions. *J. Dairy Sci.* 89: 4583–4595.

Diez-Gonzalez, F., T. R. Callaway, M. G. Kizouliz, and J. B. Russell. 1998. Grain feeding and the dissemination of acid-resistant *Escherichia coli* from cattle. *Science* 281:1666–1668.

Dodd, C. C., M. W. Sanderson, J. M. Sargeant, T. G. Nagaraja, R. D. Oberst, R. A. Smith, and D. D. Griffin. 2003. Prevalence of *Escherichia coli* O157 in cattle feeds in Midwestern feedlots. *Appl. Environ. Microbiol.* 69: 5243–5247.

Dohmen, W., F. Neijenhuis, and H. Hogeveen. 2010. Relationship between udder health and hygiene on farms with an automatic milking system. *J. Dairy Sci.* 93: 4019–4033.

Donkor, E. S., K. G. Aning, and J. Quaye. 2007. Bacterial contaminations of informally marketed raw milk in Ghana. *Ghana Med. J.* 41: 58–61.

Doyle, M. P. and M. C. Erickson. 2011. Opportunities for mitigating pathogen contamination during on-farm food production. *Int. J. Food Microbiol.* 152: 54–74.

Espina, L., M. Somolinos, S. Lorán, P. Conchello, D. García, and R. Pagán. 2011. Chemical composition of commercial citrus fruit essential oils and evaluation of their antimicrobial activity acting alone or in combined processes. *Food Control* 22: 896–902.

EC. 1996. Council directive 96/23/EC of 29 April 1996 on measures to monitor certain substances and residues thereof in live animals and animal products and repealing Directives 85/358/EEC and 86/469/EEC and decisions 89/187/EEC and 91/664/EEC. *Official Journal L* 125: 1–28.

EC. 2004a. Regulation (EC) No. 852/2004 of the European Parliament and of the Council of 29 April 2004 on hygiene of foodstuffs. *Off. J. Eur. Comm.* L139: 1–54.

EC. 2004b. Regulation (EC) No. 853/2004 of the European Parliament and of the Council of 29 April 2004 laying down specific hygiene rules for food of animal origin. *Off. J. Eur. Comm.* L139: 55–205.

EFSA. 2007. Opinion of the scientific panel on contaminants in the food chain on a request from the European Commission related to hormone residues in bovine meat and meat products. *EFSA Journal* 510: 1–62.

Ellis, K. A., G. T. Innocent, M. Mihm et al. 2007. Dairy cow cleanliness and milk quality on organic and conventional farms in the UK. *J. Dairy Res.* 74: 302–310.

Ellis-Iversen J., R. P. Smith, S. Van Winden et al. 2008. Farm practices to control *E. coli* O157 in young cattle: A randomized controlled trial. *Vet. Res.* 39: 1–12.

Ellis-Iversen J., A. J. Cook, R. P. Smith, G. C. Pritchard, and M. Nielen. 2009. Temporal patterns and risk factors for *Escherichia coli* O157 and *Campylobacter* spp. in young cattle. *J. Food Protect.* 72: 490–496.

Elmoslemany, A. M., G. P. Keefe, I. R. Dohoo, and B. M. Jayarao. 2009a. Risk factors for bacteriological quality of bulk tank milk in Prince Edward Island dairy herds. Part 1: Overall risk factors. *J. Dairy Sci.* 92: 2634–2643.

Elmoslemany, A. M., G. P. Keefe, I. R. Dohoo, and B. M. Jayarao. 2009b. Risk factors for bacteriological quality of bulk tank milk in Prince Edward Island dairy herds. Part 2: Bacteria count-specific risk factors. *J. Dairy Sci.* 92: 2644–2652.

Elmoslemany, A. M., G. P. Keefe, I. R. Dohoo, J. J. Wichtel, H. Stryhnand, and R. T. Dingwell. 2010. The association between bulk tank milk analysis for raw milk quality and on-farm management practices. *Prev. Vet. Med.* 95: 32–40.

Espejo, L. A., M. I. Endres, and J. A. Salfer. 2006. Prevalence of lameness in high-producing Holstein cows housed in freestall barns in Minnesota. *J. Dairy Sci.* 89: 3052–3058.

Espina L., M. Somolinos, S. Lorán, P. Conchello, D. García, and R. Pagán. 2011. Chemical composition of commercial citrus fruit essential oils and evaluation of their antimicrobial activity acting alone or in combined processes. *Food Control* 22: 896–902.

Evans, S. and R. Davies. 1996. Case control study of multiple-resistant *Salmonella* Typhimurium DT104 infection of cattle in Great Britain. *Vet. Rec.* 139: 557–558.

Fairchild, T. P., B. J. McArthur, J. H. Moore, and W. E. Hylton. 1982. Coliform counts in various bedding material. *J. Dairy Sci.* 65: 1029–1035.

FAOSTAT. 2010. Value of imports and exports of fresh milk. FAO Statistical yearbook. http://www.fao.org/economic/ess/ess-publications/ess-yearbook/ess-yearbook2010/en/ (accessed March 12, 2012).

Fenlon, D. R. and J. Wilson. 2000. Growth of *Escherichia coli* O157 in poorly fermented laboratory silage: A possible environmental dimension in the epidemiology of *E. coli* O157. *Lett. Appl. Microbiol.* 30: 118–121.

Fischer, J. R., T. Zhao, M. Doyle et al. 2001. Experimental and field studies of *Escherichia coli* O157:H7 in white-tailed deer. *Appl. Environ. Microbiol.* 67: 1218–1224.

Fox, E., T. O'Mahony, M. Clancy, R. Dempsey, M. O'Brien, and K. Jordan. 2009. *Listeria monocytogenes* in the Irish dairy farm environment. *J. Food Protect.* 72: 1450–1456.

Fox, J. T., B. E. Depenbusch, J. S. Drouillard, and T. G. Nagaraja. 2007. Dry-rolled or steam-flaked grain-based diets and fecal shedding of *Escherichia coli* O157 in feedlot cattle. *J. Anim. Sci.* 85: 1207–1212.

FSA. 2007. Clean Beef Cattle for Slaughter. A Guide for Producers. Food Standards Agency. http://www.food.gov.uk/multimedia/pdfs/publication/cleanbeef-saf1007.pdf (accessed April 6, 2012).

Gaggìa F., P. Mattarelli, and B. Biavati. 2010. Probiotics, and prebiotics in animal feeding for safe food production. *Int. J. Food Microbiol.* 141: S15–S28.

Galton, D. M., R. W. Adkinson, C. V. Thomas, and T. W. Smith. 1982. Effects of pre-milking udder preparation on environmental bacterial contamination of milk. *J. Dairy Sci.* 65: 1540–1543.

Galton, D. M., L. G. Petersson, and W. G. Merill. 1986. Effects of premilking udder preparations practices on bacterial counts in milk and on teats. *J. Dairy Sci.* 69: 260–266.

Galton, D. M., L. G. Petersson, W. G. Merrill, D. K. Bandler, and D. E. Shuster. 1984. Effects of premilking udder preparation on bacterial population, sediment, and iodine residue in milk. *J. Dairy Sci.* 67: 2580–2589.

Gibson G. R. and M. B. Roberfroid. 1995. Dietary modulation of the human colonic microbiota: Introducing the concept of prebiotics. *J. Nutr.* 125: 1401–1412.

Gillespie, I. A., G. K. Adak, S. J. O'Brien, and F. J. Bolton. 2003. Milkborne general outbreaks of infectious intestinal disease, England and Wales, 1992–2000. *Epidemiol. Infect.* 130: 461–468.

Glickman, L. T., P. L. McDonough, S. J. Shin, J. M. Fairbrother, R. L. LaDue, and S. E. King. 1981. Bovine salmonellosis attributed to *Salmonella* Anatum-contaminated haylage and dietary stress. *J. Am. Vet. Med. Assoc.* 178: 1268–1272.

Grauke, L. J., I. T. Kudva, J. W. Yoon, C. W. Hunt, C. J. Williams, and C. J. Hovde. 2002. Gastrointestinal tract location of *Escherichia coli* O157:H7 in ruminants. *Appl. Environ. Microbiol.* 68: 2269–2277.

Gupta, A., J. Fontana, C. Crowe et al. 2003. Emergence of multidrug-resistant *Salmonella enterica* serotype Newport infections resistant to expanded-spectrum cephalosporins in the United States. *J. Infect. Dis.* 188: 1707–1716.

Gyles, C. L. 2007. Shiga toxin-producing *Escherichia coli*: An overview. *J. Anim. Sci.* 85: E45–E62.

Ham, G. A., R. A. Stock, T. J. Klopfenstein, E. M. Larson, D. H. Shain, and R. P. Huffman. 1994. Wet corn distillers byproducts compared with dried corn distillers grains with solubles as a source of protein and energy for ruminants. *J. Anim. Sci.* 72: 3246–3257.

Hancock, D., T. Besser, J. LeJeune, M. Davis, and D. Rice. 2001. The control of VTEC in the animal reservoir. *Int. J. Food Microbiol.* 66: 71–78.

Hancock, D. D., T. E. Besser, D. H. Rice, E. D. Ebel, D. E. Herriott, and L. V. Carpenter. 1998. Multiple sources of *Escherichia coli* O157 in feedlots and dairy farms in the Northwestern USA. *Prev. Vet. Med.* 35: 11–19.

Harvey, J., K. P. Keenan, and A. Gilmour. 2007. Assessing biofilm formation by *Listeria monocytogenes* strains. *Food Microbiol.* 24: 380–392.

Hassan, L., H. O. Mohammad, and P. L. McDonough. 2001. Farm-management and milk practices associated with the presence of *Listeria monocytogenes* in New York state dairy herds. *Prev. Vet. Med.* 51: 63–73.

Hassan, L., H. O. Mohammed, P. L. McDonough, and R. N. Gonzalez. 2000. A cross-sectional study on the prevalence of *Listeria monocytogenes* and *Salmonella* in New York dairy herds. *J. Dairy Sci.* 83: 2441–2447.

Headrick, M. L., B. Timbo, K. C. Klontz, and S. B. Werner. 1997. Profile of raw milk consumers in California. *Public Health Rep.* 112: 418–422.

Headrick, M. L., S. Korangy, N. H. Bean et al. 1998. The epidemiology of raw milk-associated foodborne disease outbreaks reported in the United States, 19673 through 1992. *Am. J. Public Health* 88: 1219–1221.

Hegarty, H., M. B. O'Sullivan, J. Buckley, and C. Foley-Nolan. 2002. Continued raw milk consumption on farms: Why? *Commun. Dis. Public Health* 5: 151–156.

Heuvelink, A. E., C. van Heerwaarden, A. Zwartkruis-Nahuis et al. 2009. Two outbreaks of campylobacteriosis associated with the consumption of raw cow milks. *Int. J. Food Microbiol.* 134: 70–74.

Horchner, P. M., D. Brett, B. Gormley, I. Jenson, and A. M. Pointon. 2006. HACCP-based approach to the derivation of an on-farm food safety program for the Australian red meat industry. *Food Control* 17: 497–510.

Hovde, C. J., P. R. Austin, K. A. Cloud, C. J. Williams, and C. W. Hunt. 1999. Effect of cattle diet on *Escherichia coli* O157:H7 acid resistance. *Appl. Environ. Microbiol.* 65: 3233–3235.

Hughes, J. 2001. A system for assessing cow cleanliness. *Farm Anim. Pract.* 23: 517–524.

Huntington, G. B. 1997. Starch utilization by ruminants: From basics to the bunk. *J. Anim. Sci.* 75: 852–867.

Hutchinson, M. L., L. D. Walters, S. M. Avery, F. Munro, and A. Moore. 2005. Analyses of livestock production, waste storage, and pathogen levels and prevalences in farm manures. *Appl. Environ. Microbiol.* 71: 1231–1236.

Jacob, M. E., J. T. Fox, J. S. Drouillard, D. G. Renter, and T. G. Nagaraja. 2008. Effects of dried distillers' grain on fecal prevalence and growth of *Escherichia coli* O157 in batch culture fermentations from cattle. *Appl. Environ. Microbiol.* 74: 38–43.

Jayarao, B. M. and D. R. Henning. 2001. Prevalence of foodborne pathogens in bulk tank milk. *J. Dairy Sci.* 84: 2157–2162.

Jayarao, B. M., S. C. Donaldson, B. A. Straley, A. A. Sawant, N. V. Hegde, and J. L. Brown. 2006. A survey of foodborne pathogens in bulk tank milk and raw milk consumption among farm families in Pennsylvania. *J. Dairy Sci.* 89: 2451–2458.

Jayarao, B. M. and D. R. Wolfgang. 2003. Bulk-tank milk analysis: A useful tool for improving milk quality and herd udder health. *Vet. Clin. North Am. Food Anim. Pract.* 19: 75–92.

Johnston, A. M. 1990. Foodborne illness: Veterinary sources of foodborne illness. *Lancet* 336: 856–858.

Keene, J. E. and R. O. Elder. 2002. Isolation of shiga-toxigenic *Escherichia coli* O157 from hide surfaces and the oral cavity of finished beef feedlot cattle. *J. Am. Vet. Med. Assoc.* 220: 756–763.

Kim, S. G., E. H. Kim, C. J. Lafferty, and E. Dubovi. 2005. *Coxiella burnetii* in bulk tank milk samples, United States. *Emerg. Infect. Dis.* 11: 619–621.

Koohmaraie, M., T. M. Arthur, J. M. Bosilevac, M. Guerini, S. D. Shackelford, and T. L. Wheeler. 2005. Post-harvest interventions to reduce/eliminate pathogens in beef. *Meat Sci.* 71: 79–91.

Kousta, M., M. Mataragas, P. Skandamis, and E. H. Drosinos. 2010. Prevalence and sources of cheese contamination with pathogens at farm and processing levels. *Food Control* 21: 805–815.

Latorre, A., J. A. S. Van Kessel, J. S. Karns et al. 2009. Molecular ecology of *Listeria monocytogenes*: Evidence for a reservoir in milking equipment on a dairy farm. *Appl. Environ. Microbiol.* 75: 1315–1323.

Latorre, A., J. A. S. Van Kessel, J. S. Karns et al. 2010. Biofilm in milking equipment on a dairy farm as a potential source of bulk tank milk contamination with *Listeria monocytogenes*. *J. Dairy Sci.* 93: 2792–2802.

LeJeune, J. T., T. E. Besser, and D. D. Hancock. 2001. Cattle water troughs as reservoirs of *Escherichia coli* O157. *Appl. Environ. Microbiol.* 67: 3053–3057.

LeJeune, J. T., D. Hancock, Y. Wasteson, E. Skjerve, and A. M. Urdahl. 2006. Comparison of *E. coli* O157 and Shiga toxin-encoding genes (*stx*) prevalence between Ohio, USA and Norwegian dairy cattle. *Int. J. Food Microbiol.* 109: 19–24.

LeJeune, J. T. and P. J. Rajala-Schultz. 2009. Unpasteurized milk: A continued public health threat. *Clin. Infect. Dis.* 48: 93–100.

LeJeune, J. T. and A. N. Wetzel. 2007. Preharvest control of *Escherichia coli* O157 in cattle. *J. Anim. Sci.* 85: E73–E80.

Lira, W. M., C. Macedo, and J. M. Marin. 2004. The incidence of Shiga toxin-producing *Escherichia coli* in cattle with mastitis in Brazil. *J. Appl. Microbiol.* 97: 861–866.

Loneragan, G. H. and M. M. Brashears. 2005. Pre-harvest interventions to reduce carriage of *E. coli* O157 by harvest-ready feedlot cattle. *Meat Sci.* 71: 72–78.

Low, J. C. and Donachie, W. (1997).A review of *Listeria monocytogenes* and listeriosis. *Veterinary J.* 153(1): 9–29.

Low, J. C., I. J. McKendrick, C. McKechnie et al. 2005. Rectal carriage of enterohemorrhagic *Escherichia coli* O157 in slaughtered cattle. *Appl. Environ. Microbiol.* 71: 93–97.

Manners, J. and H. Craven. 2003. Liquid milk for the consumer. In *Encyclopedia of Food Sciences and Nutrition*, ed. B. Caballero, L. Trugo, and P. Finglas, 3947–3951. Oxford, UK: Elsevier Science Ltd.

Manninen, E., A. Marie de Passillé, J. Rushen, M. Norring, and H. Saloniemi. 2002. Preferences of dairy cows kept in unheated buildings for different kind of cubicle flooring. *Appl. Anim. Behav. Sci.* 75: 281–292.

Matthews, L., J. C. Low, D. L. Gally et al. 2006. Heterogenous shedding of *Escherichia coli* O157 in cattle and its implications for control. *Proc. Natl. Acad. Sci. U. S. A.* 103: 547–552.

Mazurek, J., E. Salehi, D. Propes et al. 2004. A multistate outbreak of *Salmonella enterica* serotype Typhimurium infection linked to raw milk consumption—Ohio, 2003. *J. Food Protect.* 67: 2165–2170.

McEvoy, J. M., A. M. Doherty, M. Finnerty et al. 2000. The relationship between hide cleanliness and bacterial numbers on beef carcasses at a commercial abattoir. *Lett. Appl. Microbiol.* 30: 390–395.

McWilliam Leitch, E. C. and C. S. Stewart. 2002. *Escherichia coli* O157 and non-O157 isolates are more susceptible to L-lactate than to D-lactate. *Appl. Environ. Microbiol.* 68: 4676–4678.

Mhone, T. A., G. Matope, and P. T. Saidi. 2011. Aerobic bacterial, coliform, *Escherichia coli* and *Staphylococcus aureus* of raw and processed milk from selected smallholder dairy farms of Zimbabwe. *Int. J. Food Microbiol.* 151: 223–228.

Millogo, V., K. S. Sjaunja, G. A. Ouédraogo, and S. Agenäs. 2010. Raw milk hygiene at farms, processing units and local markets in Burkina Faso. *Food Control* 21: 1070–1074.

Muraoka, W., C. Gay, D. Knowles, and M. Borucki. 2003. Prevalence of *Listeria monocytogenes* subtypes in bulk milk of the Pacific Northwest. *J. Food Protect.* 66: 1413–1419.

Murinda, S. E., L. T. Nguyen, S. J. Ivey et al. 2002a. Prevalence and molecular characterization of *Escherichia* coli O157:H7 in bulk tank milk and fecal samples from cull cows: A 12-month survey of dairy farms in East Tennessee. *J. Food Protect.* 65: 752–759.

Murinda, S. E., L. T. Nguyen, S. J. Ivey et al. 2002b. Molecular characterization of *Salmonella* spp. isolated from bulk tank milk and cull dairy cow fecal samples. *J. Food Protect.* 65: 1100–1105.

Murphy, M., C. Cowan, and H. Meehan. 2004. A conjoint analysis of Irish consumer preferences for farmhouse cheese. *Brit. Food J.* 106: 288–300.

Nebbia, C., A. Urbani, M. Carletti, G. Gardini, A. Balbo, D. Bertarelli, and F. Girolami. 2011. Novel strategies for tracing the exposure of meat cattle to illegal growth-promoters. *Vet. J.* 189: 34–42.

Nielsen, E. M., M. N. Skov, J. J. Madsen, J. Lodal, J. B. Jespersen, and D. L. Baggesen. 2004. Verocytotoxin-producing *Escherichia coli* in wild birds and rodents in close proximity to farms. *Appl. Environ. Microbiol.* 70: 6944–6947.

Norring, M., E. Manninen, A. M. de Passillé, J. Rushen, L. Munksgaard, and H. Saloniemi. 2008. Effects of sand and straw bedding on the lying behavior, cleanliness, and hoof and hock injuries of dairy cows. *J. Dairy Sci.* 91: 570–576.

Nørrung, B. and S. Buncic. 2008. Microbial safety of meat in the European Union. *Meat Sci.* 78: 14–24.

Norwood, D. E. and A. Gilmour. 1999. Adherence of *Listeria monocytogenes* strains to stainless steel coupons. *J. Appl. Microbiol.* 86: 576–582.

Ogden, I. D., M. MacRae, and N. J. C. Strachan. 2004. Is the prevalence and shedding concentrations of *E. coli* O157 in beef cattle in Scotland seasonal? *FEMS Microbiol. Lett.* 233: 297–300.

Oliver S. P., K. J. Boor, S. C. Murphy, and S. E. Murinda. 2009. Food safety hazards associated with consumption of raw milk. *Foodborne Pathog. Dis.* 6: 793–806.

Oliver S. P., B. M. Jayarao, and R. A. Almeida. 2005. Foodborne pathogens in milk and the dairy farm environment: Food safety and public health implications. *Foodborne Pathog. Dis.* 2: 115–129.

Oliver, S. P., D. A. Patel, T. R. Callaway, and M. E. Torrence. 2008. ASAS Centennial paper: Developments and future outlook for preharvest food safety. *J. Anim. Sci.* 87: 419–437.

Omisakin, F., M. MacRae, I. D. Ogden, and N. J. C. Strachan. 2003. Concentration and prevalence of *Escherichia coli* O157 in cattle feces at slaughter. *Appl. Environ. Microbiol.* 69: 2444–2447.

Pankey, J. W. 1989. Premilking udder hygiene. *J. Dairy Sci.* 72: 1308–1312.

Peterson, R. E., T. J. Klopfenstein, G. E. Erickson et al. 2007. Effect of *Lactobacillus acidophilus* strain NP51 on *Escherichia coli* O157:H7 fecal shedding and finishing performance in beef feedlot cattle. *J. Food Protect.* 70: 287–291.

Pickard, B. M. (undated). The case for untreated milk. A special report from the association of unpasteurised milk produces and consumers. http://naturalmilk. org/ (accessed January 25, 2012).

Reid C.-A., A. Small, S. M. Avery, and S. Buncic. 2002. Presence of food-borne pathogens on cattle hides. *Food Control* 33: 411–415.

Reig, M. and F. Toldrá. 2008. Veterinary drug residues in meat: Concerns and rapid methods for detection. *Meat Sci.* 78: 60–67.

Reinemann, D. J., G. M. V. H. Wolters, P. Billon, O. Lind, and M. D. Rasmussen. 2003. Review of practices for cleaning and sanitation of milking machines. *Bulletin* 381. International Dairy Federation, Brussels, Belgium.

Reneau, J. K., A. J. Seykora, B. J. Heins, M. I. Endres, R. J. Farnsworth, and R. F. Bey. 2005. Association between hygiene scores and somatic cell scores in dairy cattle. *J. Am. Vet. Med. Assoc.* 227: 1297–1301.

Renter, D. G., J. M. Sargeant, S. E. Hygnstorm, J. D. Hoffman, and J. R. Gillespie. 2001. *Escherichia coli* O157:H7 in free-ranging deer in Nebraska. *J. Wildl. Dis.* 37: 755–760.

Ruegg, P. L. 2003. Practical food safety interventions for dairy production. *J. Dairy Sci.* 86: E1–E9.

Ruegg, P. L. and I. R. Dohoo. 1997. A benefit to cost analysis of the effect of premilking teat hygiene on somatic cell count and intramammary infections in a commercial dairy herd. *Can. Vet. J.* 38: 632–636.

Sanaa, M., B. Poutrel, J. L. Ménard, and F. Serieys. 1993. Risk factors associated with contamination of raw milk by *L. monocytogenes* in dairy farms. *J. Dairy Sci.* 76: 2891–2898.

Sargeant, J. M., D. J. Hafer, J. R. Gillespie, R. D. Oberst, and S. J. Flood. 1999. Prevalence of *Escherichia coli* O157:H7 in white-tailed deer sharing rangeland with cattle. *J. Am. Vet. Med. Assoc.* 215: 792–794.

Scaife, H. R., D. Cowan, J. Finney, S. F. Kinghorn-Perry, and B. Crook. 2006. Wild rabbits (*Oryctolagus cuniculus*) as potential carriers of verocytotoxin-producing *Escherichia coli*. *Vet. Rec.* 159: 175–178.

Schreiner, D. A. and P. L. Ruegg. 2003. Relationship between udder and leg hygiene scores and subclinical mastitis. *J. Dairy Sci.* 86: 3460–3465.

Schrezenmeir, J. and M. de Vrese. 2001. Probiotics, prebiotics, and synbiotics— approaching a definition. *Am. J. Clin. Nutr.* 73: 361S–364S.

Shere, J. A., K. J. Bartlett, and C. W. Kaspar. 1998. Longitudinal study of *Escherichia coli* O157:H7 dissemination on four dairy farms in Wisconsin. *Appl. Environ. Microbiol.* 64: 1390–1399.

Shiferaw, B., S. Yang, P. Cieslak et al. 2000. Prevalence of high-risk food consumption and food-handling practices among adults: A multistate survey, 1996 to 1997. *J. Food Protect.* 63: 1538–1543.

Smith, D., M. Blackford, S. Younts et al. 2001. Ecological relationships between the prevalence of cattle shedding *Escherichia coli* O157:H7 and characteristics of the cattle or conditions of the feedlot pen. *J. Food Protect.* 64: 1899–1903.

Sofos, J. 2008. Challenges to meat safety in the 21st century. *Meat Sci.* 78: 3–13.

Spiegel, J. E. 2008. Making their case for raw milk. *The New York Times.* February 24. http://www.nytimes.com/2008/02/24/nyregion/nyregionspecial2/24milkct. html (accessed January 25, 2012).

Spiehs, M. J., M. H. Whitney, and G. C. Shursinm. 2002. Nutrient database for distiller's dried grains with solubles produced from new ethanol plants in Minnesota and South Dakota. *J. Anim. Sci.* 80: 2639–2645.

Spika, J. S., S. H. Waterman, G. W. Soo Hoo et al. 1987. Chloramphenicol-resistant *Salmonella* Newport traced through hamburger to dairy farms. *N. Engl. J. Med.* 316: 565–570.

Stead, S. L., H. Ashwin, S. F. Richmond et al. 2008. Evaluation and validation according to international standards of the Delvotest® SP-NT screening assay for antimicrobial drugs in milk. *Int. Dairy J.* 18: 3–11.

Steele, M. L., W. B. McNab, C. Poppe et al. 1997. Survey of Ontario bulk tank raw milk for food-borne pathogens. *J. Food Protect.* 60: 1341–1346.

Stephens, T. P., T. A. McAllister, and K. Stanford. 2009. Perineal swabs reveal effect of super shedders on the transmission of *Escherichia coli* O157:H7 in commercial feedlots. *J. Anim. Sci.* 87: 4151–4160.

Svoboda, K. P. and R. I. Greenaway. 2003. Lemon scented plants. *Int. J. Aromatherapy* 13: 23–32.

Synge, B. A., M. E. Chase-Topping, G. F. Hopkins et al. 2003. Factors influencing the shedding of verocytotoxin-producing *Escherichia coli* O157 by beef suckler cows. *Epidemiol. Infect.* 130: 301–312.

Tabe, E. S., J. Oloya, D. K. Doetkott, M. L. Bauer, P. S. Gibbs, and M. L. Khaitsa. 2008. Comparative effect of direct-fed microbials on fecal shedding of *Escherichia coli* O157:H7 and *Salmonella* in naturally infected feedlot cattle. *J. Food Protect.* 71: 539–544.

Tan, J., K. A. Janniand, and R. D. Appleman. 1993. Milking system dynamics. 2. Analysis of vacuum systems. *J. Dairy Sci.* 76: 2204–2212.

Threlfall, E. J., L. R. Ward, J. A. Frost, and G. A. Willshaw. 2000. The emergence and spread of antibiotic resistance in food-borne bacteria. *Int. J. Food Microbiol.* 62: 1–5.

Thimothe, J., K. K. Nightingale, K. Gall, V. N. Scott, and M. Wiedmann. 2004. Tracking of *Listeria monocytogenes* in smoked fish processing plants. *J. Food Protect.* 67: 328–341.

Tkalcic, S., T. Zhao, B. G. Harmon, M. P. Doyle, C. A. Brown, and P. Zhao. 2003. Fecal shedding of enterhemorrhagic *Escherichia coli* in weaned calves following treatment with probiotic *Escherichia coli*. *J. Food Protect.* 66: 1184–1189.

To, H., K. K. Htwe, N. Kako et al. 1998. Prevalence of *Coxiella burnetii* infection in dairy cattle with reproductive disorders. *J. Vet. Med. Sci.* 60: 859–861.

Tucker, C. B., D. M. Weary, and D. Fraser. 2003. Effects of three types of free-stall surfaces on preferences and stall usage by dairy cows. *J. Dairy Sci.* 86: 521–529.

USFDA. 2007. Grade "A" pasteurized milk ordinance (2007 revision) section 1 through section 7. http://www.fda.gov/Food/FoodSafety/Product-Specific Information/MilkSafety/NationalConferenceonInterstateMilkShipments NCIMSModelDocuments/PasteurizedMilkOrdinance2007/ucm063876.htm (accessed January 27, 2012).

USFDA. 2011. Steroid hormone implants used for growth in food-producing animals. U.S. Food and Drug Administration. http://www.fda.gov/AnimalVeterinary/ SafetyHealth/ProductSafetyInformation/ucm055436.htm (accessed March 12, 2012).

Van Baale, M. J., J. M. Sargeant, D. P. Gnad, B. M. DeBey, K. F. Lechtenberg, and T. G. Nagaraja. 2004. Effect of forage or grain diets with or without monensin on ruminal persistence and fecal *Escherichia coli* O157:H7 in cattle. *Appl. Environ. Microbiol.* 70: 5336–5342.

Van Donkersgoed, J., D. Hancock, D. Rogan, and A. Potter. 2005. *Escherichia coli* O157:H7 vaccine field trial in 9 feedlots in Alberta and Saskatchewan. *Can. Vet. J.* 46: 724–728.

Van Kessel, J. S., J. S. Karns, L. Gorski, B. J. McCluskey, and M. L. Perdue. 2004. Prevalence of *Salmonellae, Listeria monocytogenes*, and fecal coliforms in bulk tank milk on US dairies. *J. Dairy Sci.* 87: 2822–2830.

Van Peteguem, C. and E. Daeselaire. 2004. Residues of growth promoters. In *Handbook of Food Analysis*, ed. L. M. L. Nollet, 1037–1063, 2nd edition. New York: Marcel Dekker.

Varel, V. H., J. E. Wells, E. D. Berry et al. 2008. Odorant production and persistence of *Escherichia coli* in manure slurries from cattle fed zero, twenty, forty, or sixty percent wet distillers grains with solubles. *J. Anim. Sci.* 86: 3617–3627.

Vilar, M. J., J. L. Rodríguez-Otero, M. L. Sanjuán, F. J. Diéguez, M. Varela, and E. Yus. 2012. Implementation of HACCP to control the influence of milking equipment and cooling tank on the milk quality. *Trends Food Sci. Technol.* 23: 4–12.

Vilar, M. J., E. Yus, M. L. Sanjuán, F. J. Diéguezand, and J. L. Rodríguez-Otero. 2007. Prevalence of and risk factors for *Listeria* species on dairy farms. *J. Dairy Sci.* 90: 5083–5088.

Waak, E., W. Thamand, and M.-L. Danielsson-Tham. 2002. Prevalence and fingerprinting of *Listeria monocytogenes* strains isolated from raw whole milk in farm bulk tanks and in dairy plant receiving tanks. *Appl. Environ. Microbiol.* 68: 3366–3370.

Wallace, J. S., Cheasty, T. and K. Jones. 1997. Isolation of vero cytotoxin-producing *Escherichia coli* O157 from wild birds. *J. Appl. Microbiol.* 82: 399–404.

Ward, W. R., J. W. Hughes, W. B. Faull, P. J. Cripps, J. P. Sutherland, and J. E. Sutherst. 2002. Observational study of temperature, moisture, pH and bacteria in straw bedding, and faecal consistency, cleanliness and mastitis in cows in four dairy herds. *Vet. Rec.* 151: 199–206.

Wells, J. E., E. D. Berry, and V. H. Varel. 2005. Effects of common forage phenolic acids on *Escherichia coli* O157:H7 viability in bovine feces. *Appl. Environ. Microbiol.* 71: 7974–7979.

Wells, J. E., S. D. Shackelford, E. D. Berry et al. 2009. Prevalence and level of *Escherichia coli* O157:H7 in feces and on hides of feedlot steers fed diets with or without wet distillers grains with solubles. *J. Food Protect.* 72: 1624–1633.

Wetzel, A. N. and J. T. LeJeune. 2006. Clonal dissemination of *Escherichia coli* O157:H7 subtypes among dairy farms in Northeast Ohio. *Appl. Environ. Microbiol.* 72: 2621–2626.

Wiedmann, M. 2003. ADSA Foundation Scholar Award—An integrated science-based approach to dairy food safety: *Listeria monocytogenes* as a model system. *J. Dairy Sci.* 86: 1865–1875.

Wood, J. C., I. J. McKendrick, and G. Gettinby. 2006. Assessing the efficacy of within-animal control strategies against *E. coli* O157: A simulation study. *Prev. Vet. Med.* 74: 194–211.

Younts-Dahl, S. M., M. L. Galyean, G. H. Loneragan, N. A. Elam, and M. M. Brashears. 2004. Dietary supplementation with *Lactobacillus*- and *Propionibacterium*-based direct-fed microbials and prevalence of *Escherichia coli* O157 in beef feedlot cattle and on hides at harvest. *J. Food Protect.* 67: 889–893.

Zschöck, M., B. Kloppert, W. Wolter, H. P. Hamannand, and C. Lämmler. 2005. Pattern of enterotoxin genes *seg, seh, sei* and *sej* positive *Staphylococcus aureus* isolated from bovine mastitis. *Vet. Microbiol.* 108: 243–249.

5

Managing Risks in the Broiler
and Pig Industries

5.1 Managing Risks in the Broiler Industry

5.1.1 Introduction

Meat consumption has risen dramatically and is likely to continue to increase into the future. As society gains higher incomes, they acquire the ability to purchase meat products such as poultry and pig (Fiala, 2008). Chicken is one of the most widely accepted muscle foods in the world with high quality protein, relatively low fat content, and generally low selling price (Maurer, 2003). The United States leads the world in chicken meat production followed by China, Brazil, Mexico, and Russia (Figure 5.1a). Provided that chicken are slaughtered according to *halal* (an Arabic term meaning permitted or lawful) requirements, the meat generally has very little or no cultural or religious bindings which allows increased chicken production and consumption worldwide.

Unlike chicken, pig meats (and other carnivorous animals) are not considered halal in Muslim communities, and consumption is strictly forbidden (Nakyinsige et al., 2012).

China tops the list as the major producer of pork followed by the United States, Spain, Brazil, and Russia (Figure 5.1b). China also has the world's largest consumer market for pork; the Chinese consume about half of the world pork production annually (Amponsah et al., 2003). With an increase in meat and poultry consumption, more intensive agricultural production is adopted, leading to new challenges in microbiological hazards.

5.1.2 Main Microbiological Risks Associated with the Broiler Industry

This chapter will mainly focus on *Campylobacter* and *Salmonella*—two of the most significant foodborne pathogens in the broiler industry. The poultry discussed in this chapter focus principally on broilers, which are largely grown for meat production purposes. In the United States, salmonellosis was the most common foodborne infection reported in 2010 (8256 infections;

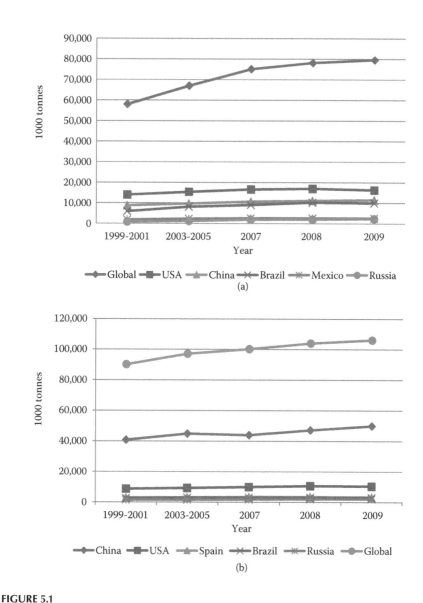

FIGURE 5.1
Top five producing countries for (a) chicken meat and (b) pig meat from 1999–2009. (From FAO, 2010. Agricultural production. FAO statistical yearbook. http://www.fao.org/economic/ess/ess-publications/ess-yearbook/ess-yearbook2010/yearbook2010-production/en/ [accessed March 16, 2010]. With permission.)

17.6 illnesses/100,000 persons) and had the largest number of hospitalizations (2290) and deaths (29) (CDC, 2011; Scallan et al., 2011). The most common serotypes were *Salmonella* Enteritidis (22%), *S.* Newport (14%), and *S.*
Typhimurium (13%). Human salmonellosis cases attributable to chicken
products were estimated at 48% (Guo et al., 2011). *Campylobacter* infections
resulted in 6365 cases (13.6 illnesses/100,000 persons) (CDC, 2011). Meanwhile,
salmonellosis is the second-ranking zoonosis in the European Union after
campylobacteriosis. Of the more than 2400 serovars of *Salmonella enterica;*
Salmonella Enteritidis is the most frequently reported serovar causing more
than 50% of the reported human *Salmonella* infections in the European Union
(EFSA, 2007). The second most reported serovar is *Salmonella* Typhimurium
being reported as being responsible for 13% of the human cases (Hald, 2008).

5.1.2.1 Campylobacter in Broilers

Poultry are considered to be a major source of human infection and provide
a transmission route for campylobacteriosis (Pearson et al., 2000). Outbreaks
associated with chicken meat have been documented in several countries
such as Denmark (Mazick et al., 2006), Australia (Black et al., 2006), and Spain
(Jimenéz et al., 2005). For every reported case of campylobacter infection,
Wheeler et al. (2001) estimated there were approximately nine others that
were not reported. There may be substantial underreporting as most individuals regard diarrhea as a transient inconvenience rather than a symptom
of disease and hence may not consult a doctor. In addition, for the reporting system to function, the general practitioner must order a stool culture,
and the laboratory must identify the etiologic agent and report the positive
results to the local or state public health institution (Rocourt et al., 2003).
Campylobacter is ubiquitous in nature and in domestic animals, and is frequently found in the environment and on many raw foods. All commercial
poultry species can carry *Campylobacter*, but the risk is greater from chicken
because of the high levels of consumption (Humphrey et al., 2007).

Findings showed that chicken meat is often contaminated with
Campylobacter spp. at the retail level, which serves as potential sources of
foodborne diseases (Hänninen et al., 2000; Suzuki and Yamamoto, 2009;
Wingstrand et al., 2006). This is supported by a sentinel surveillance system in Belgium conducted by Vellinga and Van Loock (2002), when poultry
and eggs were withdrawn in May/June 1999 due to contamination of dioxin
in the feeds. This action resulted in a 40% decline in human *Campylobacter*
infections. In Europe, the prevalence varies from 2.9% (Finland) to 71.2%
(France) (Table 5.1). Northern Europe shows lower figures compared to
southern European countries. For example, there was lower prevalence of
Campylobacter spp. in broiler flocks in Finland (2.9%; Perko-Mäkela et al., 2002),
Norway (4.8%; Hofshagen and Kruse, 2005), and Estonia (12.3%; Meremäe et
al., 2010). The number of *Campylobacter*-positive flocks was higher in Austria
(45%; Hein et al., 2003) and Germany (41.1%; Atanassova and Ring, 1999).

TABLE 5.1

Prevalence of *Campylobacter* in Broiler Chickens

Country	Number of *Campylobacter*-Positive Flocks/Total Number of Tested Flocks (%)	Reference
Austria	30/66 (45)	Hein et al., 2003
Canada	28/81 (35)	Arsenault et al., 2007
Denmark	1971/4286 (46.0)	Wedderkopp et al., 2000
Denmark	3787/8911 (42.5)	Wedderkopp et al., 2001
Estonia	163/1320 broiler chicken meat samples (12.3)	Meremäe et al., 2010
Finland	33/1132 (2.9)	Perko-Mäkela et al., 2002
France	52/73 (71.2)	Huneau-Salaün et al., 2007
Germany	209/509 poultry samples (41.1)	Atanassova and Ring, 1999
Great Britain	206/603 (34.2)	Ellis-Iversen et al., 2009
Iceland	168/1091 (15.4)	Barrios et al., 2006
Japan	(60)-Poultry	Suzuki and Yamamoto, 2009
The Netherlands	131/495 (26.3)	Bouwknegt et al., 2004
The Netherlands	153/187 (82)	Jacobs-Reitsma et al., 1994
Northern Ireland	163/388 (42)	McDowell et al., 2008
Norway	32/176 (18)	Kapperud et al., 1993
Norway	521/10,803 (4.8)	Hofshagen and Kruse, 2005
Senegal	40/70 (63)	Cardinale et al., 2004a
Sweden	77/287 (27)	Berndtson et al., 1996b
United States	28/32 (87.5)	Stern et al., 2001

Newell and Fearnley (2003) suggested that probably, the poultry industries in these northern European countries are less mature (e.g., newer facilities) and are more closely regulated than elsewhere in Europe. The number of animals per farm, climatic conditions, and distance between farms may all influence flock prevalence as well. It has also been argued that control is easier in Scandinavia because the industry is smaller and generally more economical. The winters are much harsher, hence the environmental load of *Campylobacters* is lower (Humphrey et al., 2007).

Reducing *Campylobacter* contamination of poultry products should reduce the risk of human infection. However, efforts within the slaughterhouse to reduce *Campylobacter* contamination of poultry carcasses may not be sufficient (Mead et al., 1999). In fact the slaughterhouse is an important source of contamination. For example, during scalding, contamination from hot water may spread to noncontaminated carcasses, there may be inadequate decontamination of crates, and contamination can occur during evisceration (Tsola et al., 2008). Poultry meat production in the slaughterhouse includes the following stages: slaughtering, bleeding, scalding, defeathering, evisceration, washing, chilling, and classification (Escudero-Gilete et al., 2007).

Postharvest control where *Campylobacter*-positive flocks are disposed for production of frozen poultry products and while *Campylobacter*-negative flocks are disposed for the production of fresh poultry products may not be feasible when the demand for fresh poultry exceeds the supply of *Campylobacter*-negative flocks (Wegener, 2010). Hence, control measures should start at the farm level since it is the most important primary contamination site of *Campylobacter* due to its ubiquitous presence outside the environment of the broiler houses (Jacobs-Reitsma, 2000). This is in agreement with the studies conducted by Allen et al. (2007) and Berndtson et al. (1996) who established that flocks colonized with *Campylobacter* during rearing contribute to carcass contamination during processing.

Horizontal transmission occurs when broilers are infected with *Campylobacter* spp. from the environment or other chickens. The external environment is thought to be the most important source of *Campylobacter* since *Campylobacter* is ubiquitous in farm surroundings and may come from wild and domestic animals (Humphrey et al., 2007). It is generally considered the most significant cause of *C. jejuni* infection in broiler chickens (Newell and Fearnley, 2003); hence, Wagenaar et al. (2006) suggested that primary control measures should be implemented at the farm level. In contrast, Havelaar et al. (2007) reported that the reduction of contamination at broiler farms can be efficient only in theory. It can be accomplished through improved biosecurity, but the impact of biosecurity cannot be quantified, nor can biosecurity be applied as specific control measures; it is unclear which hygienic measures and investments need to be taken (Katsma et al., 2007). Wedderkopp et al. (2001) also demonstrated that there were no significant effects on *Campylobacter* occurrence after intensive cleaning and disinfection procedures. *Campylobacter* spp. is indeed a fascinating microorganism. Although *Campylobacter* spp. may be exposed to hostile conditions (e.g., high or low temperatures) and it is sensitivity to extraintestinal conditions, its infectious potential is not compromised by these exposures. Little is known about how *Campylobacter* adapts to the hostile conditions and its ability to transmit from the food chain to consumers (Humphrey et al., 2007).

5.1.2.2 Salmonella in Broilers

Salmonella is a robust organism that can be transmitted vertically or pseudo vertically by survival of organisms on fecally contaminated eggs. The occurrence of *Salmonella* in the breeding flock can therefore result in widespread infection throughout the production system (Hensel and Neubauer, 2002). Human *Salmonella* Enteritidis cases are most commonly associated with consumption of contaminated eggs and broiler meat, while *Salmonella* Typhimurium cases are most often associated with consumption of contaminated pig, poultry, and bovine meat (EFSA, 2007). Some of the most common *Salmonella* serotypes associated with human infections in the United States are *S.* Enteritidis, *S.* Typhimurium, *S.* Newport, *S.* Javiana, and *S.* Heidelberg

(CDC, 2009). In many countries all over the world such as in Brazil (Tavechio et al., 2002), Senegal (Cardinale et al., 2004b), and Turkey (Carli et al., 2001), a wide range of different *Salmonella* serotypes were found to contaminate the broiler flocks (Table 5.2).

Surveillance of the prevalence of *Salmonella* in broiler and laying flock holdings in the United Kingdom was conducted. Sampling revealed that 41 out of 382 broiler flocks were tested positive for *Salmonella* (Snow et al., 2008) while 54 out of 454 layer flocks were found positive (Table 5.2). The most common serovar identified was *Salmonella* Enteritidis with a prevalence of 5.8%, followed by *Salmonella* Typhimurium at 1.8% of the farms (Snow et al., 2007). Improved hygiene and biosecurity measures, and vaccination of breeders and layers have considerably reduced the egg-borne transmission of *Salmonella* Enteritidis. Hence, the potential for horizontal transmission from the farm, hatchery environment, feed (Angen et al., 1996), drinkers, litter, and air (Hoover et al., 1997) need to be emphasized. Both vertical and horizontal transmissions play an important role in the contamination of flocks with *Salmonella* (Namata et al., 2009).

5.1.3 Sources and Transmission Routes of Foodborne Pathogens in Broilers

5.1.3.1 *Water and Feed*

Studies by Kapperud et al. (1993), Messens et al. (2009), Pearson et al. (1993), and Zimmer et al. (2003) showed that contaminated water is a potential source of *Campylobacter* spp. in broiler flocks. An outbreak of *C. jejuni* affecting 19 people in Southern England was reported in November 1984 to February 1985. An epidemiological investigation successfully tracked the outbreak source to the farm of origin where the farm water supply was found to be colonized with *Campylobacter* (Pearson et al., 2000). In another study by Pearson et al. (1993), they assessed the role of water as the transmission agent; by feeding *Campylobacter*-free chicks with water from the farm supply and reared under laboratory conditions, the chicks became colonized with the outbreak serotype. He and his colleagues also designed an intervention program based on water chlorination, cleaning, and disinfection of the shed's drinking system, and withdrawal of furazolidone from feed. The intervention reduces the *Campylobacter* infection from 81 to 7% in the birds. This is supported by other studies reporting that the addition of sanitizers reduces the risk of *Campylobacter* infection (Ellis-Iversen et al., 2009; Evans and Sayers, 2000; Jeffrey et al., 2001; Kapperud et al., 1993). Chaveerach et al. (2004) have shown that acidified drinking water can prevent *Campylobacter* spread in broiler flocks. In contrast, a study by Stern et al. (2002) reported that chlorination of drinking water was not effective in decreasing colonization by *Campylobacter* spp. under commercial broiler production practices in the United States. Lindblom et al. (1986) also argued that water contamination

TABLE 5.2

Prevalence of *Salmonella* in Broiler Chickens

Country	Number of *Salmonella*-Positive Flocks/Total Number of Tested Flocks (%)	Profile of Most Common Serotypes (%)	Reference
France	93/519 (17.9)	S. Typhimurium (23.7) S. Enteritidis (21.5) S. Infantis (8.6)	Huneau-Salaün et al., 2009
Spain	728/2678 broiler samples (from 44 broiler flocks)	S. Enteritidis (66) S. Virchow (13.0) S. Hadar (12.0) S. Seftenberg (2.6)	Marin et al., 2009
United Kingdom	41/382 broiler flocks (10.7) 54/454 layer flocks (11.9)	S. Enteritidis (5.8) S. Typhimurium (1.8)	Snow et al., 2007, 2008
Japan	10/70 broiler chicken samples (14)	S. Infantis	Asai et al., 2006
Denmark	27/1174 batches (2.3)	Not reported	EFSA Denmark, 2006
Finland	< 1%	Not reported (Finnish *Salmonella* Control Programme)	EFSA Finland, 2006
Sweden	8/196 (4)	National *Salmonella* Control Programme	EFSA Sweden, 2006
Senegal	20/70 (28.6)	S. Hadar (12.8) S. Brancaster (8.6)	Cardinale et al., 2004b
Brazil	994 poultry flock samples/4581 samples (foodstuffs, poultry flocks, environment, sewage, water, animal feed) (21.7)	S. Enteritidis (11.4) S. Seftenberg (2.1) S. Hadar (0.57) S. Agona (0.39)	Tavechio et al., 2002
Turkey	151/814 samples (18.6)	S. Enteritidis (81.5) S. Angona (7.6) S. Thompson (10.1) S. Sarajane (0.8) S. Enteritidis (19.8)	Carli et al., 2001
	490/8911 (5.5)	S. Typhimurium (17.9) S. Infantis (17.5)	Wedderkopp et al., 2001
Saudi Arabia	1052/25,759 poultry and environment samples (4)	S. Enteritidis (isolated from 98.8% of all samples) S. Virchow (isolated from 57.8% of all samples) S. Paratyphi B var. Java (isolated from 57.7% of all samples)	Al-Nakhli et al., 1999
Canada	27/635 samples (4.3)	S. Hadar (2) S. Heidelberg (0.5)	Chambers et al., 1998

follows rather than precedes colonization of a flock, suggesting that this is a consequence of contamination of water lines by organisms excreted from the birds while other studies determined that water source is only a low-risk factor (Jacobs-Reitsma et al., 1994; Humphrey et al., 1993). Berndtson et al. (1996b) reported that feed is not a risk factor in *Campylobacter* infection. The dry conditions of feed are considered lethal to *C. jejuni*.

Renwick et al. (1992) found that *Salmonella* infection in broiler flocks were significantly associated with the type of drinker used, where the odds of isolating *Salmonella* from water samples taken from metal troughs or plastic bell drinkers was significantly higher than nipple drinkers. Drinkers with open designs such as troughs and bell drinkers are more prone to fecal contamination than closed designs such as nipple drinkers.

Jones and Richardson (2004) noted that feed is a potential source of *Salmonella* into broiler production. It was determined that both feed ingredients and dust were the major culprit in *Salmonella* contamination in feed mills. For example, soybean meal is often contaminated and approximately 30% of all consignments imported into Sweden are positive for *Salmonella* (Lewerin et al., 2005). Other high-risk ingredients are rapeseed meal, corn gluten meal, and fish meal (Lewerin et al., 2005). Alvarez et al. (2003) isolated 32 different serovars from Spanish feeds while studies from Jones and Richardson (2004) determined that 19 of 451 pelleted feed samples (4.21%) were contaminated with *Salmonella*. All poultry feed must be heat treated at a minimum of 75°C. However, inappropriate heat treatment during pelleting, condensation in the coolers and storage bins of the mill, recontamination after heat treatment, and poor hygienic practices of feed mills are important risk factors (Lewerin et al., 2005). Dust within feed mills is considered a major source of contamination (Jones and Richardson, 2004). According to Davies and Hinton (2000), persistent contamination in feed mills usually occurs in the pellet coolers. Pellet coolers pulling large volumes of air and dust obtained within the cooler would have a greater likelihood of contamination than dust collected from other areas (Jones and Richardson, 2004). There is also insufficient time to close the mill and to conduct a complete decontamination series of cleaning and fumigation while maintaining a feed supply for customers (Davies, 2005). *Salmonella* serovar Enteritidis obtained from feed samples and egg contents from a layer farm showed similar pulsed-field gel electrophoresis patterns, suggesting that the contamination in the eggs and farm was linked to the feed (Shirota et al., 2001). Serovars found in feed mills were also detected in the birds fed with the feed during rearing and slaughter (Corry et al., 2002).

5.1.3.2 Cleanliness of Animals

Proper cleaning of commercial poultry production units is important for a successful biosecurity program. The effectiveness of a cleaning regimen depends on the techniques used, choice of disinfectants, organic load, building design, and wall and floor materials (Rathgeber et al., 2009; Wales et al., 2006). For example, wet cleaning which was thought to be more effective than dry cleaning in the removal of debris was in fact only moderately effective in the studies by Wales et al. (2006). Most animal houses have cracks and crevices in floors, walls, and ceilings, which allow entry of wildlife vectors.

Carryover from a previous flock (Petersen and Wedderkopp, 2001) and repeated infections with certain *Campylobacter* clones in broiler houses may contribute to transmission of *Campylobacter* in broiler chickens (Wedderkopp et al., 2003). However, numerous studies reported that the carryover from one flock to a subsequent flock is a relatively infrequent event (Pearson et al. 1996; Van de Giessen et al., 1992). This is because cleaning and disinfection of poultry houses are generally sufficient to reduce the risk of *Campylobacter* transmission from an infected flock to a subsequent flock housed in the same rearing hall (Shreeve et al., 2002). The correct method of cleaning and disinfection are important, because Huneau-Salaün et al. (2007) found out that the risk of a broiler flock being colonized with *Campylobacter* was increased when the first disinfection of the poultry house was performed by the farmer instead of a hygiene specialist. Although farmers used methods and disinfectants similar to those employed by hygiene specialists, probably the application method was not as thorough as the professional contractors. Ellis-Iversen et al. (2009) also reported that if the previous flock in the house had been *Campylobacter* positive, the first batch of the following flock was also more likely to be infected. However, they argued that this association could be due to a persistent risk practice or source of *Campylobacter* on the farm rather than a direct carryover from the previous flock. In a study by Messens et al. (2009), farms with four and five broiler houses were all found negative for *Campylobacter*. Her findings show that the presence of other broilers on the same farm did not give rise to a more frequent colonization among broilers in different houses. This may be that good biosecurity is practiced on the farms and help to prevent or delay flock colonization (Gibbens et al., 2001). Nevertheless, biosecurity breeches occur. Every broiler house on a farm is entered by production staff or the farmer at least once and usually three to four times daily. In addition, maintenance workers and other authorized visitors may break the barrier as well (Refrégier-Petton et al., 2001). Thus, in the life of the broilers (approximately 42 days), this means that there are approximately 50 to 150 occasions on which the barrier is broken by human traffic; each providing an opportunity for the entrance of *Campylobacter* (Wagenaar et al., 2008).

Poultry houses without cemented floors were associated with an elevated risk of *Campylobacter* infection. Removing used litter from a cemented floor is easier than from an earth floor. Moreover, a cemented floor is easier to dry than an earth floor and a dry condition is lethal to *Campylobacter* (Cardinale et al., 2004a). When side fans are used for ventilation, there is less risk of *Campylobacter* infection due to its accessibility for cleaning and disinfection as compared to roof fans (Gibbens et al., 2001).

Transport crates, which are used to load and transport birds between farm and lairage, may be contaminated with *Salmonella* and *Campylobacter*, despite having been cleaned at the factory (Corry et al., 2002). Transport crates can be heavily soiled by feces, uric acid, litter material, and feathers from the birds. Birds may produce more excreta possibly exacerbated by increased stress levels during transportation (Rostagno, 2009; Tinker et al., 2005).

5.1.3.3 Wild/Domestic Animals and Insects

Environmental contamination during rearing of poultry is the most likely source of *Campylobacter* infection (Berndtson et al., 1996a; Van de Giessen et al., 1998). Wild bird droppings found on or near broiler houses in four farms were screened for *Campylobacter*, and between 4 to 11% of 57 samples were found positive for *C. jejuni* (Craven et al., 2000). Studer et al. (1999) found 68% of 231 soil samples within four chicken farms were positive for *C. jejuni* or *C. coli*, suggesting that the surrounding soil might constitute as potential reservoir of *Campylobacter*. Flies may be an important vector since flies had the potential to transmit *C. jejuni* from outside animals to chickens in the broiler house as they may take up *Campylobacter* as they forage on fresh animal feces. A total of 8.2% of 49 flies caught outside a broiler house were tested positive for *C. jejuni*. Furthermore, the number of flies entering the broiler house increases as the need for ventilation air increases. It is estimated that about 30,000 flies passed through the ventilation system into the broiler house in one broiler cycle during the summer season (Hald et al., 2004). A reduction of *Campylobacter* spp. colonization in broilers in farms will subsequently reduce the infectious microbial loads entering the abattoirs and processing plants, thus reducing human exposure to campylobacteriosis through chicken meat (Ellis-Iversen et al., 2009).

The ingestion of beetles contaminated with *Salmonella* could be a vector for transmission to broilers (Roche et al., 2009). *Typhaea stercorea* is determined as a potential carrier of *S.* Infantis between successive broiler flocks (Hald et al., 1998). Angen et al. (1996) who conducted a review of Denmark's database of 7108 broiler flocks did not find an association between the presence of beetles and *Salmonella*. This is in agreement with Gradel and Rattenborg (2003) whose results indicate that beetles do not play a role in persistent *Salmonella* infections.

5.1.3.4 Age and Flock Size

Increasing age and flock size were shown to contribute to a higher prevalence of *Campylobacter* infection in broiler flocks as well (Barrios et al., 2006; Berndtson et al., 1996a). Evans and Sayers (2000) found 40% of 100 broiler flocks in Great Britain were infected by the time the chicks were 4 weeks old and >90% by 7 weeks. The hypothesis suggested by Bouwknegt et al. (2004) for the age effects in *Campylobacter* infection might be due to the introduction of solid feeds to chicks, which alters the intestinal ecology. A longer duration in the broiler house meant more chances for *Campylobacter* to be introduced from the house environment (Barrios et al., 2006) and additional time before slaughter would also allow the cecal-colony concentrations to become detectable (Stern et al., 2001). Larger flocks also require more water, feed, litter, air, and working hours all representing potential sources of infection (Berndtson et al., 1996a). While McDowell et al. (2008) indicated that the increasing odds of infection observed with increasing age is due to the cumulative risk of introduction of infection over time from the environment. Jacobs-Reitsma et al. (1994) demonstrated that there is a higher rate of infection in summertime compared to the winter period. The effects for this seasonal pattern are unknown but Nylen et al. (2002) suggested that this may reflect levels of environmental contamination. This argument is supported by the fact that poultry houses have more ventilation in the summer; hence, potentially increasing the contact with the outside environment (Wagenaar et al., 2008).

Lahellac et al. (1986) conducted an epidemiological survey in 10 poultry operations and found that the main source of contamination was the initial *Salmonella* colonization in the chicks. This study is in agreement with Kim et al. (2007) who revealed that 1-day old chicks infected with *Salmonella* are an important risk factor to a farm. Positive chicks can spread the infection through their feces. The boxes in which they arrive may be a potential source of infection as well. The study revealed that S. Enteritidis was consistently found in the integrated broiler operation: the breeder farm, hatcheries, broiler farms, and chicken slaughterhouse. The PFGE pattern was related in the whole broiler chain. Samples from day-old chicks showed high *Salmonella* prevalence at 34.2% (Marin et al., 2009). Younger broilers also appeared to be at higher risk of S. Enteritidis infection (Hald, 2008) and if infection occurs in newly hatched chicks, it could remain as a persistent infection (Van Immerseel et al., 2004).

5.1.3.5 Flock Thinning Practices

Flock thinning practices or batch depletion in broiler houses increase the prevalence of *Campylobacter* spp. infected broilers in the flocks (Slader et al. 2002; Wedderkopp et al., 2000). The introduction of *Campylobacter* occurred during the thinning practice, followed by the entire flock within a week. When chickens are caught for slaughter, the intensive traffic to and from the broiler house during catching may expose the broiler house to *Campylobacter* spp. Thus, the area available to the remaining broiler may become newly contaminated with *Campylobacter* spp. (Hald et al., 2001). The entry of catching personnel and equipment during the process of partial depopulation also serve as a potential source of infection for the remaining birds in the house (McDowell et al., 2008).

According to Newell and Fearnley (2003), colonization of *Campylobacter* in broiler chicken flocks spreads most commonly by horizontal transmission. Previous studies showed that when specific biosecurity measures such as twice weekly replenishment of boot dip disinfectant; appropriate location of ventilation fans and daily sanitization of water supply (Gibbens et al., 2001); and monitoring of staff's hygiene and changing of clothes and boots in each broiler house and cemented poultry house floors (Cardinale et al., 2004a) are followed, *Campylobacter* infection in broilers was reduced by over 50% (Van de Giessen et al., 1998).

5.1.4 Intervention Strategies

The increasing demand for chicken meat has led to a higher degree of intensification and advances in breeding, specially formulated diets, disease control, and management practices that enabled the broiler industry to produce a chicken weighing 1.8 kg in 6–7 weeks (Maurer, 2003). The intensification (increasing population density within a reduced space) is often associated with higher burdens and risks of infection with animal pathogens (Graham et al., 2008). Bailey et al. (2001) reported that it is likely that no single intervention or *Salmonella* control strategy will consistently reduce *Salmonella* on the farm. But the basis for successful control of *Salmonella* infections in poultry farms are good farming and hygienic practices (including feed, birds management, cleaning and disinfection, and control of rodents), as well as testing and removal of positive flocks from production (EFSA, 2004). Collard et al. (2007) also mentioned that in order to manage the risk to human health, it is essential to tackle the problem at the farm level to reduce the cross-contamination occurring throughout the food chain process. For laying hens, risk factors are large flock size (Mollenhorst et al., 2005; Namata et al., 2008), probably because a higher flock size increases the number of susceptible birds and large-sized poultry houses are often linked to egg-packing plants by means of a common egg conveyor. Murase et al. (2001) demonstrated that *Salmonella* may spread from a contaminated poultry house to another through common egg conveyors.

5.1.4.1 Biosecurity Measures

In the poultry industry, biosecurity is defined as the implementation of a set of measures that minimizes the risks of introduction and spread of disease agents in animals. The three principles of biosecurity are segregation, cleaning, and disinfection. Segregation is the creation and maintenance of physical or virtual barriers to limit the potential opportunities for infected animals and contaminated materials to enter an uninfected site. Segregation, if properly applied, is the most effective way to prevent infection. Secondly, cleaning will help to remove most of the contaminating pathogens and disease agents. This is followed by disinfection, which will inactivate microorganisms that are present on surfaces that have been thoroughly cleaned (FAO, 2008).

The operation of *Salmonella*-free pig and poultry units shows how biosecurity on farms can directly contribute to an improvement in the health status of livestock (Collins and Wall, 2004). Sanitary practices such as the use of a footbath may be a risk factor of *Salmonella* contamination. The footbath has to be clean to prevent the introduction and spread of pathogens within the house. It should be placed inside the anteroom in the house or covered to avoid inactivation of the disinfectant or dilution by rainwater, and renewed at least twice weekly (Aury et al., 2010). Lewerin et al. (2005) also suggested that the use of a footbath is not sufficient for decontamination purposes, and a change of footwear is required. Farmers and visitors can bring in pathogens on their clothes and shoes, and therefore be potential vectors (Cardinale et al., 2004b). Previous studies showed that *Campylobacter* infection in broilers was reduced by over 50% when specified biosecurity measures were followed, for example, twice weekly replenishment of boot dip disinfectant, location of ventilation fans and daily sanitization of the water supply (Gibbens et al., 2001), staff hygiene and changing of clothes and boots for each broiler house, and mixed animal farm and cemented poultry house floors (Cardinale et al., 2004a). In fact, Johnsen et al. (2006) determined that the farm with the poorest biosecurity routines had broilers that became infected earliest.

Some of the generally practiced biosecurity measures (Doyle and Erickson, 2006; Racicot et al., 2011) in broiler farms are:

- Respecting "contaminated" and clean areas (e.g., an anteroom separated in two with the area closest to the door giving access to the birds being considered "clean")
- Changing into barn-specific boots
- Wearing barn-specific overalls
- Hand washing before entering barn
- Hand washing when exiting barn
- Disinfecting footwear using a footbath
- Restricting and minimizing traffic onto farm, into houses, and between flocks; compulsory registration of visitors
- Cleaning housing and equipment between flocks including disinfection of the water supply system
- Removing all manure between flocks
- Minimizing exposure of equipment, feed, and flock to wild animals (e.g., birds, rodents)
- Cleaning and disinfecting transport crates and vehicles after every use

According to Wedderkopp et al. (2001), the Danish *Salmonella* Control Programme found that cleaning samples have to be taken after the detection of *S.* Enteritidis or *S.* Typhimurium in a flock, and the broiler producers must plan an extended period of empty housing of at least 14 days. The Danish producers are genuinely motivated to improve the cleaning and disinfection of the broiler houses, as *Salmonella*-free broiler flocks obtain a higher price at slaughter. The Danish government and the European Union also compensate owners of destroyed breeding stock for their losses (Wegener et al., 2003). Improvements in breeding flocks and feed, improved cleaning, and disinfection of houses have reduced *Salmonella* to a low level in the commercial broiler production in the United Kingdom. There is still the need to reduce the prevalence of *Campylobacter* and possible introduction of viral diseases such as the avian influenza in farms (Davies, 2005).

5.1.4.2 Feed and Water

Pelleting has been recommended as an effective method for feed decontamination. During the processing of feed pellets, feed are conditioned, followed by pelleting or expanding. Feed meal is usually introduced into a conditioner where steam is added to raise the temperature. A temperature of 176°F (80°C) in the conditioner followed by pelleting is sufficient to kill *Salmonella* (EFSA, 2008). Pelleting usually involves temperatures between 158°F and 194°F (McCapes et al., 1989). In addition to feed treatments, the application of HACCP principles and Good Hygiene Practices (GHP)/Good Manufacturing Practices (GMP) systems should be applied in the feed chain (i.e., from feed ingredients to feed processing, storage, and transport). Chemical treatment of feedstuffs has been utilized to reduce existing levels of pathogens in feed and minimize recontamination (Ricke, 2005). A number of compounds such as acetic acid, propionic acid, buffered propionate, citric acid, formaldehyde, formic acid, lactic acid, and phosphoric acid had been considered (EFSA, 2008; Ricke, 2005). The antibacterial mechanisms of organic acids are primarily due to their ability to disrupt pH gradients and intracellular pH regulations. By lowering the pH, the organic acid molecules will be undissociated and, because in this state they carry no net charge, they can cross through the lipid membranes of bacterial cells with ease. Once inside the cell, the acids reduce the near-neutral pH value of the cytoplasm, which results in potential membrane damage and/or anion accumulation (Cherrington et al., 1991; Ricke, 2003; Van Immerseel et al., 2006). Synergistic combinations of formic and propionic acids (Anderson et al., 2001) or formaldehyde and propionic acid appears to reduce *Salmonella* shedding in laying flocks (Thompson and Hinton, 1997).

Treatments of water supplies with organic acids have shown a reduction of *Salmonella* in broilers. Byrd et al. (2001) added 0.5% of acetic, lactic, or formic acid into drinking water during an 8-hour pretransport feed withdrawal period. Both formic and lactic acids reduced the number of *Salmonella* present. Supplementation with sodium nitrate and chlorate product in drinking water (Byrd et al., 2003; Jung et al., 2003) also showed beneficial results in reducing *Salmonella* in broilers. The inclusion of organic acid in the drinking water during the pretransport feed withdrawal period may be beneficial in reducing microbial load in broilers prior to processing. However, the drawback in using acidified drinking water is the corrosion affecting galvanized pipes, joints, and nipple drinkers (Wales et al., 2010).

5.1.4.3 Diet

Diet formulations may affect the prevalence of pathogens in poultry. For example, chickens receiving plant protein-based feed has significantly less colonization of *C. jejuni* in their ceca than birds receiving other types of feed (Udayamputhoor et al., 2003). Broilers fed with a maize-based diet showed significant reduction of *Salmonella* in spleen, liver, and ceca compared to broilers receiving a wheat-rye-based diet (Teirlynck et al., 2009). *Salmonella* colonization in poultry may be influenced by grain type and particle size (Huang et al., 2006). Feeding whole wheat compared to ground wheat in pelleted diets decreased intestinal colonization of *Salmonella* in broilers (Bjerrum et al., 2005), while feeding mash decreased *Salmonella* in gizzards and ceca of broilers compared to feeding pellets (Huang et al., 2006). Results obtained by Santos et al. (2008) found that broilers fed whole or coarsely ground grains have lower *Salmonella* populations. Using triticale as a feed ingredient compared to corn results in improved body weight (2.75 vs. 2.64 kg) and reduced *Salmonella* colonization.

5.1.4.4 Competitive Exclusion

Competitive exclusion (CE) bacteria are nonpathogenic bacteria typically found in the gastrointestinal tract of animals and may be composed of a single specific strain or several strains or species of bacteria. CE is introduced early in the birds' life so that the CE bacteria are preferentially established in the gastrointestinal system to become competitive or antagonistic to pathogens (Doyle and Erickson, 2006). To be effective, CE treatment should be administered before birds' are exposed to *Salmonella*. Hence, by spraying hatching eggs or chicks in hatching trays are generally superior to adding CE to feed or drinking water (Cox et al., 1992; Mead, 2000). Moreover, administration via feed or water may result in variable protection among the flock due to differential rates of feeding or drinking by the birds. The number of viable CE bacteria may also be reduced before the bird consumes the feed or drinks the water (Doyle and Erickson, 2006).

5.1.4.5 Vaccination

Vaccination involves the external activation of the bird's own immune system to limit pathogen colonization. Birds are exposed to vaccines specific to targeted microorganisms. The birds then can develop specific antibodies against the pathogens as part of its immune responses. There are two types of vaccines that can be used: (1) inactivated or killed bacterial cells, and (2) live attenuated bacterial strain (strains devoid of significant side effects but still capable of inducing immunity) (Sirsat et al., 2009). However, vaccination of broilers with a live *Salmonella* vaccine have been shown to be more effective in causing immunity than killed or inactivated cells (Mastroeni et al.,

2001; Totton et al., 2012; Van Immerseel et al., 2009). Belgium implemented a *Salmonella* eradication program in poultry in accordance with the European legislation 2160/2003 (EU, 2003) in which a vaccination program is implemented in breeders and layers but not in broilers because of the short shelf-life expectancy of broilers (42 days) (EFSA, 2004). Studies based on killed vaccines tend to focus on laying breeds since an immune response following vaccination with a killed vaccine takes time. Layers are better samples compared to broilers as the latter tend to be slaughtered before immunity can develop (Totton et al., 2012; Van Immerseel et al., 2009). In many countries (United Kingdom, United States, the Netherlands, Germany, Italy, Spain, France, Argentina, Brazil, Thailand), vaccination of breeder flocks and commercial layer flocks using approved live or inactivated *Salmonella* vaccines is permitted (Breytenbach, 2004). Vaccination is an effective control in commercial applications, but vaccines can be extremely specific to a particular bacterium and may not be effective against diverse pathogens. For *Salmonella*, this could be problematic due to the wide range of serotypes (Sirsat et al., 2009).

5.1.4.6 Education and Training

The education of poultry farmers to improve general hygiene and enhance disease prevention has the advantage to address biosecurity against infectious poultry foodborne diseases. Norway reported a positive effect from its farmer education programs (Hofshagen and Kruse, 2005) but Iceland failed to observe any effect (Wagenaar et al., 2008).

The foodborne outbreaks occurring in the meat production chain have been occurring for years. Most of the meat-borne outbreaks were due to improper food handling practices and consumption of undercooked meat. However, the majority of pathogenic bacteria that can spread at slaughter by cross-contamination were traced back to production on the farm rather than originating from the slaughter plant. *Escherichia coli* O157:H7, *Salmonella* spp., and *Campylobacter* are a few of the prominent pathogens to affect the meat chain. The hazards occurring in the primary production should be adequately controlled as required by EU Directive 852/2004. Effective risk assessment and Good Farming Practices should be rigorously adopted to mitigate risks at the farm level.

5.2 Managing Food Safety Risks in Pigs

5.2.1 Introduction

Globally, an estimated 80.3 million illnesses and 155,000 foodborne-related deaths were attributed to nontyphoidal *Salmonella* (Majowicz et al., 2010).

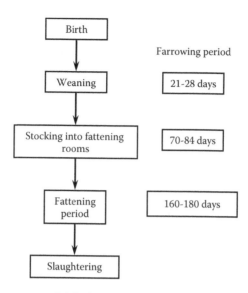

FIGURE 5.2
Farrow-finish pig herd production.

Around 10–20% of human *Salmonella* infections in the European Union are attributable to pigs (EFSA, 2010). Meanwhile in the United States, the number of human salmonellosis cases directly attributable to pork or pork products was estimated at <1% (Guo et al., 2011). The pig production discussed in this chapter focuses principally on farrower to finisher, with stock largely grown for slaughtering purposes (Figure 5.2).

Salmonella surveillance and monitoring programs have been implemented in several countries in Europe and control programs at slaughter in North America (Funk and Gebreyes, 2004). According to Rostagno and Callaway (in press), even though *Salmonella* has been identified throughout the pig production chain, there has been an increasing focus at the farm level. However, the control of *Salmonella* is very complex at the preharvest stages due to the numerous potential sources of contamination. Poor monitoring of farms, increasing farm size, dietary changes, and lack of a specific *Salmonella* control program may have contributed to the prevalence and dissemination of *Salmonella* infection (Blaha, 2001). Hence, primary production units have received special attention in control and surveillance programs within the European Union (EU). In the EU Regulations No. 2003/99/EC and 2160/2003, the European Commission implemented a *Salmonella* surveillance program in different livestock species (Directive 2003/99/EC; Regulation EC No. 2160/2003). The surveillance programs for pigs were carried out using serologic detection of antibodies using either serum or meat juice samples (Alban et al., 2002; BPEX, 2002; Davies et al., 2003). In the United Kingdom, the British Pig Executive (BPEX) and Department for Environment, Food and Rural Affairs (DEFRA) and the Food Standards Agency (FSA), introduced the Zoonoses Action Plan (ZAP) *Salmonella* Program to monitor *Salmonella* prevalence in 2002 (BPEX, 2002) and was replaced by the Zoonosis National Control Plan (ZNCP) for *Salmonella* in pigs in 2008 (BPEX, 2009). Serological surveillance for *Salmonella* was based on the presence of antibodies as an indicator of the presence of *Salmonella* (Abrahantes et al., 2009). The Scandinavian countries Sweden, Norway, and Finland have a very low prevalence of *Salmonella*. These countries implemented preharvest surveillance programs and an eradication strategy program (Bengtsson et al., 2009; Huttunen et al., 2006; Jore et al., 2009).

In contrast, in the U.S. no preharvest surveillance but decontamination procedures are carried out in abattoirs (Alban et al., 2012). Alban and Stärk (2005) revealed that in order to reduce *Salmonella* prevalence in a given pig herd from level 3 (*Salmonella* prevalence >10%) to level 1 (<1% prevalence), it would cost US$ 6.34–7.92 per pig (for a slaughter pig herd). A farm to table approach may not be practical in the US due to differences in production systems, industry structure, and regulatory organization. *Salmonella* in the U.S. pork chain is controlled by the Pathogen Reduction: Hazard Analysis and Critical Control Point (HACCP) systems which established performance standards at slaughter and processing plants. This program has resulted in decreased contamination of product with *Salmonellae* (FSIS, 1996, 2011).

Salmonella-infected pigs usually remain asymptomatic and may excrete *Salmonella* spp. in feces and harbor the bacteria in the tonsils, intestines, and gut-associated lymphoid tissue (GALT) (Fedorka-Cray et al., 2000). The most commonly isolated nontyphoidal serotypes in pigs and pork are *Salmonella* Typhimurium including variant 5 (formerly variant Copenhagen) and *Salmonella* Derby (Boyen et al., 2008; EFSA, 2006; Gebreyes et al., 2004; Valdezate et al., 2005). *Salmonella* Typhimurium is the predominant serotype isolated from humans in Europe, and pigs are probably the most important source of infection in these countries (Boyen et al., 2008). *Salmonella* carrier pigs entering the slaughterhouse seem to be the most common source of carcass and meat product contamination (Beloeil et al., 2004). *Salmonella* prevalence in pig farms or pork carcasses can range from nondetected to 63% (Table 5.3). This huge variance across countries reflects the different dynamics of *Salmonella* infection, differences between countries with different pig densities and environmental temperatures, different technologies in husbandry and slaughter of pigs, the use of different pig breeds (Swanenburg et al., 2001b), and different *Salmonella* control programs in farms and abattoirs. What we would also like to demonstrate here are the various types of samples (i.e., as lymph nodes, cecal, fecal, carcass, and the environment) colonized by different *Salmonella* serotypes. *Salmonella* prevalence in pigs differ a lot depending on which part of the pig is sampled. Not all different samples of the pig will become available for human consumption, but collecting more than one type of samples per pig showed that *Salmonella* can be found in almost the whole pig (Swanenburg et al., 2001b). *Salmonella* includes more than 2500 different serotypes (Dunkley et al., 2009; Foley and Lynne, 2008).

TABLE 5.3

Prevalence of *Salmonella* in the Pig

Country	Number of *Salmonella*-Positive Samples/Total Number of Samples (%)	Predominant Serovars	Reference
Brazil	Environmental, lymph nodes and carcass surfaces' samples 487/1258 (39)	S. Typhimurium, S. Panama, S. Seftenberg, S. Derby, and S. Mbandaka	Kich et al., 2011
Spain (free-range pigs)	Lymph nodes samples 43/804 (5.3)	S. Anatum, S. Typhimurium, S. Muenchen	Gómez-Laguna et al., 2011
Reunion Island, France	Fecal samples from 24/60 herds (40)	S. Typhimurium, S. Derby	Cardinale et al., 2010
Alberta and Saskatchewan, Canada	Fecal samples 407/1143 (35.6)	S. Derby, S. Typhimurium var. Copenhagen, S. Putten	Wilkins et al., 2010
Thailand	Fecal samples 122/194 (63)	S. Rissen; S. Typhimurium, S. Stanley, and S. Weltevreden	Dorn-in et al., 2009
New Zealand	Swab samples from pig carcasses 0/100 (0)	None detected	Wong et al., 2009
Belgium	96/584 (16)	Not reported	Delhalle et al., 2008
Saskatchewan, Canada	Cecal samples 29/232 (12.5); Ileocecal lymph node (ICLN) samples 12/232 (5.2)	S. Derby, S. Enteritidis, S. California in cecal samples; S. Typhimurium var. Copenhagen, S. Derby, S. Enteritidis in ICLN samples	Mainar-Jaime et al., 2008
Ethiopia	Pig carcasses 120/278 (43)	S. Hadar, S. Eastbourne, S. Saintpaul	Aragaw et al., 2007
Spain (fattening units)	Pooled fecal samples 290/2320 (12.5)	S. Typhimurium, S. Rissen, S. Derby	García-Feliz et al., 2007
United States	Fecal samples 46/934 (4.9)	S. Agona, S. Derby, S. Schwarzengrund, S. Typhimurium, and S. Seftenberg	Bahnson et al., 2006
Mexico	Pork samples 172/296	S. Meleagridis, S. Havana, S. Agona, S. Anatum	Zaidi et al., 2006
United States	Fecal samples 105/895 (11.7) Lymph node samples 133/895 (14.9)	S. Derby, S. Typhimurium, S. Typhimurium (Copenhagen), S. Anatum	Bahnson et al., 2005

Continued

TABLE 5.3 (*Continued*)

Prevalence of *Salmonella* in the Pig

Country	Number of *Salmonella*-Positive Samples/Total Number of Samples (%)	Predominant Serovars	Reference
Vietnam	Pork samples 95/136 (69.9)	*S.* Derby, *S.* Weltevreden, *S.* London	Phan et al., 2005
Great Britain	Cecal samples 578/2507 (23)	*S.* Typhimurium, *S.* Derby	Davies et al., 2004
Illinois, USA	75/3534 (2)	*S.* Derby, *S.* Agona, *S.* Worthington, and *S.* Uganda	Barber et al., 2002
The Netherlands	Serum samples 33/183 (18) Carcass swab samples 3/213 (1.4) Liver swabs 14/150 (9.3) Tongue swabs 14/151 (9.3) Rectal contents 21/82 (25.6) Mesenterial lymph nodes 14/150 (9.3) Tonsils 35/179 (19.6) Slaughterhouse environment 87/377 (23)	*S.* Typhimurium, *S.* Panama, *S.* S14,5,12:d:2ef, *S.* Derby, *S.* Livingstone, *S.* Mbanka	Swanenburg et al., 2001b

5.2.2 On-Farm Risk Factors

Although *Salmonella* has been identified throughout the pork chain, there has been an increasing focus on the preharvest phase (i.e., on-farm stage) (Rostagno and Callaway, in press). By identifying on-farm risk factors, one can establish adequate and cost-effective intervention strategies at the herd level in order to reduce *Salmonella* prevalence in slaughter pigs and hence producing safe pork (Dorn-in et al., 2009). A number of risk factors exist at the farm level—which includes feed type and feeding practices, purchased pigs, herd size, hygienic practices, floor type, contact between pigs, and transport and lairage.

5.2.2.1 Feed and Water

Animal feed is crucial to the growth and development of livestock. It may also be the source of infection at the farm level (in this case—*Salmonella* spp.) especially in countries with low *Salmonella* prevalence. Industrial pig feed and feed produced on-farm utilized products and co-products from primary production (agriculture) and from food processing industries. Cereals such as barley, wheat, corn, and rye form the basic energy sources while major protein sources were derived from soybean meal, rapeseed, and sunflower meals (Binter et al., 2011). Major proteinaceous feed materials are at risk of *Salmonella* contamination and may introduce *Salmonella* into feed mills and pig farms (EFSA, 2010).

Wierup and Häggblom (2010) found that out of the 38 *Salmonella* serovars isolated from vegetable feed ingredients (28) and from feed mills (10), 30 of the serovars had also been isolated from human cases of salmonellosis in Sweden between 1997–2008. Contamination of feedstuffs was also associated with *Salmonella* infection in pigs (Davies et al., 2004; Molla et al., 2010; Österberg et al., 2006). For example, in 2003, a feedborne outbreak of *Salmonella* Cubana occurred in Sweden as a result of contamination in the feed plant. *Salmonella* Cubana was later detected in 49 out of 77 pig farms that received the contaminated feed (Österberg et al., 2006). Feeding pelleted feed was found to increase risk of *Salmonella* infection in pigs (Farzan et al., 2006; García-Feliz et al., 2009; Wilkins et al., 2010). In an *in vitro* study, Hedemann et al. (2005) found that *Salmonella* adhered less to the intestinal tissue of pigs fed the nonpelleted diets than to those fed pelleted diets. It was suggested that pigs fed pelleted diets secrete mucins that are capable of binding *Salmonella enterica*, allowing colonization. Pigs fed coarse nonpelleted feed also showed higher undissociated lactic acid in gastric content and increased concentration of butyric acid in the cecum and colon compared to pigs fed fine nonpelleted feed (Mikkelsen et al., 2004). Feeding coarse ground meal feed to pigs affect the physicochemical and microbial properties of content in pigs' stomach, which decreases the survival of *Salmonella*. Hence, pigs fed a nonpelleted diet were better protected against *Salmonella* infections than pigs fed a pelleted diet.

The risk for *Salmonella* shedding was increased when dry feed (versus wet feed) was provided during the fattening period (Beloeil et al., 2004; Farzan et al., 2006; Lo Fo Wong et al., 2004) and America (Bahnson et al., 2006). Why is this so? In liquid feed with fermented by-products, the natural fermentation process increases the growth of lactic acid bacteria (Prohászka et al., 1990; Van der Wolf et al., 2001a). The ingredient present in the liquid feed such as whey or distiller's grain is more likely to support the growth of lactic acid bacteria by acting in a prebiotic manner (Farzan et al., 2006). The low pH (between 4.0 and 4.5) and the presence of organic acids and large numbers of lactic acid producing bacteria might explain this protective effect (Lo Fo Wong et al., 2002). Missotten et al. (2009) also screened for potential lactic acid bacteria to produce fermented liquid feed. *Lactobacillus johnsonii, L. salivarius*, and *L. plantarum* were found to be very effective for the production of fermented liquid feed and showed high antimicrobial activity against *Salmonella* spp.

During primary production of feed, contamination is possible through spreading of contaminated fertilizers (e.g., slurry, manure, waste sludge) on the fields and pasture (EFSA, 2008). Outdoor reared pigs with outdoor feeding arrangements are very attractive to wild birds, which may move between livestock farms. Feed delivered into troughs, ad-lib feeders, or hoppers may also attract rodents. Pig feeding systems with troughs, hoppers, and pipes are particularly difficult to clean and disinfect effectively because of inaccessible surfaces and pooling of wash water (EFSA, 2008). Jensen et al. (2004) and Barber et al. (2002) also detected *Salmonella* in water samples collected from water bowls, nipple-waterers, faucets, and troughs. Feder et al. (2001) also reported that *Salmonella* were detected in 63% (15 of 24) of the water samples collected from water bowls or mud holes. In a study by Rho et al. (2001), the researchers found that pathogens were not detected from the water tanks of six pig farms. However, *Salmonella* spp. was found in the water nipples of all farms. This may be either due to contamination during water transferring from water tank to nipples in pig pens or from direct contamination of nipples by the pigs.

5.2.2.2 Purchased Pigs

In countries with high *Salmonella* prevalence in pigs, infected pigs can be the dominant route of transmission (EFSA, 2010). Purchasing pigs from infected herds is a risk factor for introduction of *Salmonella* (Zheng et al., 2007). For example, Berends et al. (1996) demonstrated that *Salmonella* imported from the breeding farm probably contributed to 1–10% of all infections that take place during the finishing period. Pigs purchased from more than three suppliers had three times higher odds to test seropositive than pigs in herds which bred their own replacement stock or recruited from a maximum of three supplier herds (Lo Fo Wong et al., 2004).

5.2.2.3 Herd Size

There may be an association between herd size and the detection or prevalence of *Salmonella*. For example, García-Feliz et al. (2009) found that the risk of *Salmonella* increased when the number of slaughtered pigs per year was ≥3500. Christensen et al. (2002) also found that larger herd size (>2600) was associated with higher risk of *Salmonella*. Farzan et al. (2006) found no association between herd size and *Salmonella* shedding in pigs, while Van der Wolf et al. (2001a) found that small- to moderate-herd size (<800 pigs) were associated with a higher *Salmonella* seroprevalence compared to larger herds. The researchers speculated that the large farms were probably more hygiene-conscious than smaller farms. In addition to herd size, the number of barns, rooms per barn and pens per room, the pig densities in the barn, room and pen levels, size of production site, number of workers, nursery pigs or gilts may also be important when considering the herd size (Farzan et al., 2006; García-Feliz et al., 2009).

Van der Wolf et al. (2001c) found a high prevalence of *Salmonella* in finisher pigs reared in the free-range system. In the free-range system, each pig has to have straw as bedding material and access to an outside area except for weaned pigs. Hence, a plausible explanation for the higher prevalence of *Salmonella* in free-range finishers is the fact that regular pigs are kept on slatted floors and the free-range pigs were kept on solid floors with straw bedding. The solid floors could result in an increased contact with feces in these systems resulting in an increased fecal–oral contamination route. Organic and free-range pigs have contact with the external environment, which may constitute a risk of infection either through direct contact with wildlife and other nonproduction animals or their droppings. Contaminated water and pig "wallows" (Funk and Gebreyes, 2004; Callaway et al., 2005) can contribute to the higher prevalence. In free-range pigs, other infections such as toxoplasmosis may occur due to the increased contact with the environment.

5.2.2.4 Wild and Domestic Animals

Wildlife especially rodents had been found to carry and spread *Salmonella* on farms (Wales et al., 2009). Barber et al. (2002) found a correlation between a farm having a high prevalence of shedding *Salmonella* and a high abundance in cats (11.5%), rodents (9%), flies (16%), and bird feces (3%). *Salmonella* isolates from cat feces were also more likely to have *Salmonella* isolated from intestinal contents of rodents and from the environmental deposits of bird feces. There is a possible role for cats in serving as a source for *Salmonella* transmission due to their unhindered movement around a farm and even between farms. Contrary, there was a lower *Salmonella* prevalence in herds where a cat could enter the pig units (Nollet et al., 2004). This is in line with a study by Veling et al. (2002) in dairy herds. The authors ascribed the lower prevalence to the role of cats as predator of rodents and birds carrying or excreting *Salmonella*. On the other hand, the study by Jensen et al. (2004) did not find *Salmonella* in rodents and birds trapped in the surrounding environment of an outdoor pig farm. This was interesting as it was hypothesized a higher incidence of *Salmonella* may be due to increased exposure of pigs to the surrounding environment, particularly contact with wildlife. A total of 2377 samples from 110 species of migratory birds in Sweden were analyzed for *Salmonella* infections, where only one isolate was found positive (Hernandez et al., 2003). Hernandez et al. (2003) raised an important point where by finding infected birds close to a livestock farm does not prove that transmission occurred from the birds to the animals.

5.2.2.5 Cleaning and Disinfection

It is generally accepted that improved biosecurity measures with all in/all out management and strict cleaning and disinfection protocols reduces potential infections and *Salmonella* contamination within the farm environment and pig population. Schmidt et al. (2004) and Roesler et al. (2005) reported that cleaning and disinfection effectively reduces the amount of *Salmonella* in lairage pens. Contrary to this, Lo Fo Wong et al. (2004) found that compartments that were never disinfected after pressure washing as part of the all in-all out regimen were associated with a lower *Salmonella* seroprevalence than herds that sometimes or always used disinfectants. In addition, Poljak et al. (2008) found *Salmonella* shedding was positively associated with an increased frequency of disinfection and washing with cold water. The question why the *Salmonella* prevalence in herds that do not use disinfectants is lower than herds that do is interesting. The researchers speculate that farmers that use disinfectants clean less adequately with the idea that any remaining microbes will be dealt with by the disinfectant. However, unclean surfaces and organic matter may render the disinfectant ineffective. The occurrence of residual environmental *Salmonella* contamination of the floor and pens in finishing units also increased the risk of *Salmonella* shedding in the next batch of growing pigs (Beloeil et al., 2004). Rough floor surfaces are usually more contaminated than smooth surfaces, and this may result in the occurrence of residual *Salmonella* (Madec et al., 1999). Cleaning and disinfection of postweaning accommodation and more generally on pig farms are usually insufficient or inconsistent to remove pre-existing *Salmonella* contamination (Wales et al., 2009), so residual *Salmonella* in weaner and grower accommodation constitutes one of the major sources in the pig farm. This negates the benefits of *Salmonella*-negative incoming stock (Dahl, 2008). *Salmonella* spp. are also able to produce biofilms on a variety of surfaces to form an extracellular matrix that contribute to long-term survival and increase resistance to antimicrobial stress (White et al., 2006). The decreased ability of disinfectant agents to reach deep areas in biofilm (Stewart et al., 2001) might leave *Salmonella* protected.

5.2.2.6 Types and Design of Farm Structure

Types of flooring and surface will influence pig contact with fecal material. In a study by Davies et al. (1997), *Salmonella* infection was lower in pigs housed on slatted flooring than in those housed on solid floors with open-flush gutters. The type of flooring in the finishing unit too will influence the prevalence of *Salmonella* (e.g., prevalence in ascending order: fully slatted floor <50–99% slatted < less than 50% slatted; Nollet et al., 2004). Feces from pigs housed on slatted floors immediately flows away in the manure pit thereby reducing pigs' exposure to infected feces. On the other hand, van der Wolf et al. (2001a) found no risk associated to different types of floor. The risk for a batch to shed *Salmonella* increased when the pit under the slatted floor in the farrowing room was not emptied after removing the previous batch of pigs. A frequency of removal of sow dung of less than once per day during the lactation period also increased the risk of finishing-room contamination (Beloeil et al., 2004). Small et al. (2003) showed that the *in vitro* survival rates of *Salmonella* Kedougou were higher in bedding material than for hide, concrete, or metal.

5.2.2.7 Pen Separation

Pigs which were able to have snout contact with pigs in neighboring pens showed higher seropositivity for *Salmonella* compared to pigs in which such contact was prevented (Lo Fo Wong et al., 2004). Proux et al. (2001) conducted a study to determine *Salmonella* infection by nose-to-nose contact. Five *Salmonella*-inoculated pigs and four noninoculated "contact" pigs were housed in adjacent pens to allow nose-to-nose contact. The noninoculated "contact" pigs' fecal samples were found to be bacteriologically positive after 5 weeks. Pen separation may prevent fecal spread between adjacent pens, hence reducing the spread of infection.

5.2.2.8 Transport and Lairage

Pigs can be infected during the transport or waiting period in lairage before slaughtering. Pigs are usually held from 2 to 8 hours in the lairage before slaughter; lairage and the slaughterhouse environment are probably the major source for *Salmonella* infections prior to slaughter (Hurd et al., 2001; Swanenburg et al., 2001a). Lázaro et al. (1997) also recovered 20% *Salmonella*-positive samples from the holding pen of a pig slaughterhouse in Brazil. Swanenburg et al. (2001a) found that almost the whole lairage was contaminated with *Salmonella* when pigs were present. This will have a significant effect on the number of *Salmonella*-infected pigs at slaughter. Exploring the environment is the normal behavioral pattern of pigs; this includes sniffing, rooting, and drinking, all of which provides ample opportunity for the pigs to be infected with *Salmonella*. Swabs taken from trucks, lairage, and slaughterline before the pigs were delivered showed 4.3% (3/70), 80% (64/80), and 16.7% (4/27) samples were positive for *Salmonella* spp. (Mannion et al., 2012). There was also a build-up of *Salmonella* populations throughout the slaughter week as indicated by Boughton et al. (2007). On day 1 (Monday—start of slaughter week), the researchers isolated *Salmonella* spp. in 6% (11 out of 179) floor swabs. This was followed by day 2 (Thursday—end of slaughter week), when lairage pens were subjected to high-pressure cold water washing between batches of pigs, 44% (79 out of 180) floor swabs were positive for *Salmonella* spp. Cleaning and disinfection was successful in reducing the prevalence of *S. enterica* from lairage pens (from 20–100% to 0–15%) but was not able to consistently reduce the prevalence of *S. enterica* in pigs held in those cleaned pens. Transport and lairage are important reservoirs of *Salmonella* spp. and has the potential to infect pigs and cross-contaminate on the pig surface leading to carcass contamination.

5.2.3 Intervention Strategies

Intervention strategies can be taken during the preharvest (in the herd) or postharvest (at the slaughterhouse) stages or by a combination of the two. At the preharvest level, there are three main strategies (EFSA, 2010) to reduce the prevalence of *Salmonella* on the farm:

  i. Ensure breeder pigs are *Salmonella*-free
 ii. *Salmonella*-free feedstuffs
iii. Preventing infection from external sources (e.g., birds and rodents)

5.2.3.1 Salmonella-Free Breeder Pigs

Control of *Salmonella* in breeding herds may be relevant and effective in countries with a very low or negligible preharvest prevalence of *Salmonella*. However, it is not feasible in countries with a higher prevalence (Alban et al., 2012). In Scandinavian countries—Finland, Norway, and Sweden where prevalence of *Salmonella* was between 0 to 1.8%—breeding herds are usually monitored and bacteriologically sampled. If positive samples were found, control measures were taken (EFSA, 2009). Rather than trying to achieve *Salmonella*-free breeder pigs, it might be more practical to monitor the prevalence and type of *Salmonella* spp. present in the breeders. Pig herds can be assigned a *Salmonella* status that has to be reported to the buyer during trading. The herds can be allocated into three main categories such as: negative for *Salmonella* (Status A), positive for *S.* Typhimurium, *S.* Derby or *S.* Infantis (Status C), and positive for other types of *Salmonella* (Status B; Alban et al., 2012). The three types of serovars were selected since it was found that herds infected with these serovars remained infected for up to several years (based on Danish pig herds data; Baptista et al., 2009). Similarly, other countries with different prevalent strains can monitor their herds based on their existing data.

5.2.3.2 Feed

Heat treatment is generally recognized as the most effective decontamination method. The purpose of pelleting and heat treatment process is to reduce *Salmonella* contamination of compound feed. Although *Salmonella* is seldom isolated from pelleted (heat treated) feed at the feed mills, feed can become recontaminated with *Salmonella* during transport, storage or in the feeding system at the farm (Lo Fo Wong et al., 2004). Where GHP/GMP are in place, the risk of recontamination is minimized. In cases where heat treatment is inappropriate (e.g., pelleted feed), the feed ingredients or compound feed can be treated with blends of organic acids. The chemical treatment has a residual protective effect in feed, which helps to reduce recontamination (EFSA, 2008). In a study conducted by Creus et al. (2007), commercial pelleted feed added with 0.6% formic acid plus 0.6% lactic acid was found to reduce *Salmonella* prevalence, while diets containing 0.4% lactic acid plus 0.4% formic acid or 0.8% formic acid resulted in lower fecal excretion of *Salmonella*. In the EU, Regulation (EC) No. 183/2005 indicates that "Feed business operators shall put in place, implement and maintain a permanent written procedure or procedures based on the HACCP principles."

Lo Fo Wong et al. (2004) also investigated 359 European fattening-pig herds and found that pigs fed nonpelleted (dry or wet) feed had a lower risk of *Salmonella* seropositivity compared to pigs fed pelleted feed. This may be attributed to the structure (e.g., coarseness of feed) and composition of nonpelleted feed. Coarsely ground grain may not be digested well as compared to finely ground pelleted feed, hence some of the carbohydrates of coarsely ground grain will ferment in the large intestine. This results in the forming of volatile fatty acids, reduction of pH, and creates a hostile environment for *Salmonella*. In another previous study by van de Wolf et al. (2001c), herds fed fermented liquid feed (FLF) had a lower *Salmonella* prevalence. Similarly, the protective effect of FLF is probably due to the large amounts of organic acids and lactic acid producing bacteria present in these feeds. The organic acids are mostly lactic and acetic acid. Alternatively, organic acids can be added to drinking water (van der Wolf et al. 2001b). Contamination during primary production of feed can be decreased by composting fertilizer, ploughing in after spreading fertilizer, and by increasing the time between spreading of fertilizer and animal grazing or crop harvesting (EFSA, 2008).

5.2.3.3 Cleaning and Disinfection

The following points should be followed to reduce contamination (Madec et al., 1999):

  i. Removal of slurry from the pit below the slatted floor
 ii. Damping to be started soon after pig removal
iii. Prolonged damping
 iv. Thorough washing and disinfecting soon after washing
  v. Attention to the recommended dosage of disinfectant

Care must be taken when washing pen floors. Power-washing may cause splashing of contaminated material on to the feeders. Increasing the water spray pressure increases the generation of aerosols, and the aerosol droplets were generally smaller and may have remained suspended for longer periods. It is possible that lowering the pressure may limit the spread of contamination (Gibson et al., 1999). Provision of frequent clean bedding may provide welfare benefits in terms of pen cleanliness since straw bedding helped to keep pens and pigs cleaner (59% versus 77% clean floor area for solid floor without straw and solid floor with straw) (Spoolder et al., 2000).

5.2.3.4 Biosecurity Measures on the Farm

In pig production, biosecurity is defined as the protection of a pig herd from the introduction and spreading of infectious agents (viral, bacterial, fungal, or parasitic) (Amass and Clark, 1999).

Most biosecurity protocols on pig farms require that personnel, visitors, and veterinarians disinfect their boots before entering the facilities and when moving between groups of pigs of different ages or health status. Boot baths are used to disinfect boot surfaces in an attempt to prevent or reduce the mechanical transmission of pathogens among groups of pigs. Using boot baths at the entrance of the pig house reduces the *Salmonella* seroprevalence (Barber et al., 2002; Rajić et al., 2005). In a study on farms in Alberta, Canada, *Salmonella* was recovered from a high percentage of workers' boots (34/88), which emphasizes the need for proper disinfection of farm apparel (Rajić et al., 2005).

Amass et al. (2000) suggested that scrubbing visible manure from boots enhances removal of significant numbers of bacteria. However, simply walking through a boot bath will not reduce bacterial counts. Some possible cleaning and disinfection strategies for boots as suggested by Amass et al. (2000) are:

i. Provide designated boots for a specific area within the farm (e.g., farrowing house).

ii. Alternatively, a farm can establish boot stations which consist of a wash area for cleaning off manure and a bath of disinfectant containing spare boots. At the boot station, personnel could remove contaminated boots, clean and scrub them in the wash area, and then place the cleaned boots in the tub of disinfectant. The personnel can then put on the spare boots that had been previously soaked in the disinfectant.

iii. This will ensure that the disinfectant remains fairly free of manure and contaminants since only clean boots are added into the tub.

Most farms also have a rule that visitors must be free from exposure to pigs for 24–48 hours before entry. Human traffic within pig farms should be kept to a minimum. Pig producers can evaluate the effectiveness of current biosecurity programs by recording information regarding movements of people, animals, feed, and equipment to, from, and within their production facilities. Records can be used to inform managers of biosecurity risks or breaches and to identify the likely source of disease introduction or contamination (Amass and Clark, 1999). For example, a log sheet for signing in and out and foot baths at facility entrance and exit are useful and prevent irregular entries (Ojha and Kostrzynska, 2007).

A biosecurity scoring system was developed and considered in one of the following categories (Pinto and Urcelay, 2003):

   i. Location and isolation—To prevent entry of infectious agents in a pig farm. Includes components such as close to other farms (less than 1.5 miles), workers in contact with other pig farm workers, or keeping pigs at home;

  ii. Internal risks—Includes all hazards within farms such as pest, presence of domestic animals, disposal of dead animals, quarantine new arrivals, disease records, fallowing period;

 iii. Moveable risks—Includes elements such as feed, waste management, use of water, and personnel not moving between sections; and

  iv. Nonmovable risks—Includes elements that are fixed within the farm (e.g., equipment not shared by different sections, changing facilities, clean and disinfect trucks, use of wheel bath for trucks at entrance of bath, etc.).

5.2.3.5 HACCP in the Slaughterhouse

In 1996, the United States Food Safety and Inspection Service (FSIS) introduced the pathogen reduction and HACCP systems to control food safety hazards in fresh meat processing, including pork (FSIS, 1996). The objective of the system was to reduce the pathogen numbers over time as a result of the introduction of HACCP into meat plants. Meanwhile, in 2002, the European Union introduced the HACCP system for fresh meat slaughter, which includes pigs (EC, 2001). It is also important to note that Good Manufacturing Practices and Good Hygiene Practices are the foundation for the successful implementation of HACCP.

Interventions such as scalding, singeing, evisceration, and possibly hot water washing and chilling could be used as CCPs in the European Union pig HACCP system (Pearce et al., 2004). Scalding was found to reduce *Salmonella* from 31% to 1% in carcasses (Pearce et al., 2004). Decontamination of the final carcass with acetic or lactic acid is used in the United States (Hurd et al., 2001) but decontamination of carcasses using organic acids is not allowed in the EU (Regulation No. 853/2004). Use of acids will probably only be accepted for disinfection of tools (Alban and Stärk, 2005). In beef, washing the animal is beneficial for the control of *Salmonella*, especially if the hide is contaminated (Arthur et al., 2007). Similarly, washing and spraying pigs before they are slaughtered could further reduce the level of skin contamination (Letellier et al., 2009).

According to Bollaerts et al. (2010) and Hurd et al. (2008), interventions at primary production had minimal effect to reduce *Salmonella*. Alban and Stärk (2005) and Van der Gaag et al. (2004) found that the slaughtering stage appeared to be the most important stage in the supply chain to reduce the prevalence of *Salmonella*-contaminated carcasses. Improvements in the following processing area are beneficial:

- Increased singeing efficacy.
- Reduced probability of contamination and cross-contamination at degutting. This can be achieved by using automated procedures for degutting and disinfection of tools to prevent cross-contamination.
- Reduced probability of contamination during handling. This can be achieved through improved hygiene such as scalding the knives, washing hands, visual meat inspection and proper disinfection, and improved precision at carcass splitting (Alban and Stärk, 2005; Swanenburg et al., 2001b).

In practice, the interventions that should be implemented and their cost-effectiveness ultimately depend heavily on the farmer and risk profile of the farm. Each farm is specific and may have their own weak points (e.g., nearness to other livestock farms, reduced biosecurity measures) in preventing or controlling *Salmonella* on-farm. Hence, in addition to preharvest efforts, all steps from stable to table need to be considered.

5.2.4 *Salmonella* spp. Outbreaks Traced to Pork

Pig and pig meat products are recognized as important sources of human salmonellosis. For example, in Germany, five large outbreaks were reported from 2001 to 2005 (Jansen et al., 2007). In one of the outbreaks that occurred in summer 2001, traceback investigations indicated that contamination of pork meat occurred early in the rearing production chain. Based on the PFGE patterns of the case patients, the investigators were able to link the *Salmonella* strain back to a farm (Buchholz et al., 2005). Meanwhile, in 2008, an outbreak of *Salmonella* Typhimurium occurred in Norway, Sweden, and Denmark. In Denmark, a total of 37 cases were identified. Four patients died (all older than 75 years old) and suffered from underlying illnesses, and it remains unclear to what degree the *Salmonella* infection contributed as a cause of death. In Norway, 10 cases were identified while 4 cases were detected in Sweden. The outbreak strain was traced back to a cutting plant and samples from a sow herd (Bruun et al., 2009).

References

Abrahantes, J. C., K. Bollaerts, M. Aerts, V. Ogunsanya, and Y. Van der Stede. 2009. *Salmonella* serosurveillance: Different statistical methods to categorise pig herds based on serological data. *Prev. Vet. Med.* 89: 59–66.

Alban, L., F. M. Baptista, V. Møgelmose et al. 2012. *Salmonella* surveillance and control for finisher pigs and pork in Denmark—A case study. *Food Res. Int.* 45: 656–665.

Alban, L. and K. D. C. Stärk. 2005. Where should the effort be put to reduce the *Salmonella* prevalence in the slaughtered swine carcass effectively? *Prev. Vet. Med.* 68: 63–79.

Alban, L., H. Stege, and J. Dahl. 2002. The new classification system for slaughter-pig herds in the Danish *Salmonella* surveillance-and-control program. *Prev. Vet. Med.* 53: 133–146.

Allen, V. M., S. A. Bull, J. E. L. Corry et al. 2007. *Campylobacter* spp. contamination of chicken carcasses during processing in relation to flock colonization. *Int. J. Food Microbiol.* 113: 54–61.

Al-Nakhli, H. M., Z. H. Al-Ogaily, and T. J. Nassar. 1999. Representative *Salmonella* serovars isolated from poultry and poultry environments in Saudi Arabia. *Reveu scientifique et technique de L—Office international des épizooties* 18: 700–709.

Alvarez, J., S. Porwollik, I. Laconcha et al. 2003. Detection of a *Salmonella enterica* serovar California strain spreading in Spanish feed mills and genetic characterization with DNA microarrays. *Appl. Environ. Microbiol.* 69: 7531–7534.

Amass, S. F. and L. K. Clark. 1999. Biosecurity considerations for pork production units. *Swine Health Prod.* 7: 217–228.

Amass, S. F., B. D. Vyverberg, and D. Ragland. 2000. Evaluating the efficacy of boot baths in biosecurity protocols. *Swine Health Prod.* 8: 169–173.

Amponsah, W. A., X. D. Qin, and X. Peng. 2003. China as a potential market for US pork exports. *Rev. Agric. Econ.* 25: 259–269.

Anderson, K. E., B. W. Sheldon, and K. Richardson. 2001. Effect of Termin-8® compound on the productivity of brown egg laying chickens and environmental microbial populations. *Poultry Sci.* Suppl. 80: 14.

Angen, Ø., M. N. Skov, M. Chriél, J. F. Agger, and M. Bisgaard. 1996. A retrospective study on *Salmonella* infection in Danish broiler flocks. *Prev. Vet. Med.* 26: 223–237.

Aragaw, K., B. Molla, A. Muckle et al. 2007. The characterization of *Salmonella* serovars isolated from apparently healthy slaughtered pigs at Addis Ababa abattoir, Ethiopia. *Prev. Vet. Med.* 82: 252–261.

Arsenault, J., A. Letellier, S. Quessy, V. Normand, and M. Boulianne. 2007. Prevalence and risk factors for *Salmonella* spp. and *Campylobacter* spp. caecal colonization in broiler chicken and turkey flocks slaughtered in Quebec, Canada. *Prev. Vet. Med.* 81: 250–264.

Arthur, T. M., J. M. Bosilevac, D. M. Brichta-Harhay et al. 2007. Effects of a minimal hide wash cabinet on the levels and prevalence of *Escherichia coli* O157:H7 and *Salmonella* on the hides of beef cattle at slaughter. *J. Food Prot.* 70: 1076–1079.

Asai, T., M. Itagaki, Y. Shiroki et al. 2006. Antimicrobial resistance types and genes in *Salmonella enterica* Infantis isolates from retail raw chicken meat broiler chickens on farms. *J. Food Prot.* 69: 214–216.

Atanassova, V. and C. Ring. 1999. Prevalence of *Campylobacter* spp. in poultry and poultry meat in Germany. *Int. J. Food Microbiol.* 51: 187–190.

Aury, K., M. Chemcaly, L. Petetin et al. 2010. Prevalence and risk factors for *Salmonella enterica* subsp. *enterica* contamination in French breeding and fattening turkey flocks at the end of the rearing period. *Prev. Vet. Med.* 94: 84–93.

Bahnson, P. B., P. J. Fedorka-Cray, S. R. Ladely, and N. E. Mateus-Pinilla. 2006. Herd-level risk factors for *Salmonella enterica* subsp. *enterica* in U.S. market pigs. *Prev. Vet. Med.* 76: 249–262.

Bahnson, P. B., J.-Y. Kim, R. M. Weigel, G. Y. Miller, and H. F. Troutt. 2005. Associations between on-farm and slaughter plant detection of *Salmonella* in market-weight pigs. *J. Food Prot.* 68: 246–250.

Bailey, J. S., N. J. Stern, P. J. Fedorka-Cray et al. 2001. Sources and movement of *Salmonella* through integrated poultry operations: A multistate epidemiological investigation. *J. Food Prot.* 64: 1690–1697.

Baptista, F. M., L. Alban, A. K. Ersbøll, and L. R. Nielsen. 2009. Factors affecting persistence of high *Salmonella* serology on Danish pig herds. *Prev. Vet. Med.* 92: 301–308.

Barber, D. A., P. B. Bahnson, R. Isaacson, C. J. Jones, and R. M. Weigel. 2002. Distribution of *Salmonella* in swine production ecosystems. *J. Food Prot.* 65: 1861–1868.

Barrios, P. R., J. Reiersen, R. Lowman et al. 2006. Risk factors for *Campylobacter* spp. colonization in broiler flocks in broiler. *Prev. Vet. Med.* 74: 264–278.

Beloeil, P.-A., P. Fravalo, C. Fablet et al. 2004. Risk factors for *Salmonella enterica* subsp. *enterica* shedding by market-age pigs in French farrow-to-finish herds. *Prev. Vet. Med.* 63: 103–120.

Bengtsson, B., U. Carlsson, E. Chenais et al. 2009. *Surveillance and Control Programs: Domestic and Wild Animals in Sweden.* In, Carlsson, U. and Elvander, M. (eds.), National Veterinary Institute, Uppsala, pp. 1–72.

Berends, B. R., H. A. P. Urlings, J. M. A. Snijders, and F. van Knapen. 1996. Identification and quantification of risk factors in animal management and transport regarding *Salmonella* spp. in pigs. *Int. J. Food Microbiol.* 30: 37–53.

Berndtson, E., M.-L. Danielsson-Tham, and A. Engvall. 1996a. *Campylobacter* incidence on a chicken farm and the spread of *Campylobacter* during the slaughter process. *Int. J. Food Microbiol.* 32: 35–47.

Berndtson, E., U. Emanuelson, A. Engvall, and M.-L. Danielsson-Tham. 1996b. A 1-year epidemiological study of *Campylobacters* in 18 Swedish chicken farms. *Prev. Vet. Med.* 26: 167–185.

Binter, C., J. M. Straver, and P. Häggblom. 2011. Transmission and control of *Salmonella* in the pig feed chain: A conceptual model. *Int. J. Food Microbiol.* 145: S7–S17.

Bjerrum, L., K. Pedersen, and R. M. Engberg. 2005. The influence of whole wheat feeding on *Salmonella* infection and gut flora composition in broilers. *Avian Dis.* 49: 9–15.

Black, A. P., M. D. Kirk, and G. Millard. 2006. *Campylobacter* outbreak due to chicken consumption at an Australian Capital Territory restaurant. *Comm. Dis. Intelligence* 30: 373–377.

Blaha, T. 2001. Pre-harvest food safety as integral part of quality assurance systems in the pork chain from "stable to table". *Int. J. Food Microbiol.* 44: 7–13.

Bollaerts, K., W. Messens, M. Aerts et al. 2010. Evaluation of scenarios for reducing human salmonellosis through household consumption of fresh minced pork meat. *Risk Anal.* 30: 853–865.

Boughton, C., J. Egan, G. Kelly, B. Markey, and N. Leonard. 2007. Quantitative examination of *Salmonella* spp. in the lairage environment of a pig abattoir. *Foodborne Pathog. Dis.* 4: 26–32.

Bouwknegt, M., A. W. Van de Giessen, W. D. C. Dam-Deisz, A. H. Havelaar, N. J. D. Nagelkerke, and A. M. Henken. 2004. Risk factors for the presence of *Campylobacter* spp. in Dutch broiler flocks. *Prev. Vet. Med.* 62: 35–49.

Boyen, F., F. Haesebrouck, D. Maes, F. Van Immerseel, R. Ducatelle, and F. Pasmans. 2008. Non-typhoidal *Salmonella* infections in pigs: A closer look at epidemiology, pathogenesis and control. *Vet. Microbiol.* 130: 1–19.

BPEX. 2002. ZAP Salmonella. A zoonoses action plan for the British Pig Industry. Available at: http://www.bpex-zncp.org.uk/resources/000/241/150/ZAP_Control_Update_-_April_2002.pdf (accessed February 16, 2012).

BPEX. 2009. BPEX Annual technical report 2008-2009. Available at: http://www.bpex.org.uk/downloads/298874/293090/BPEX%20Annual%20Technical%20Report%2008-09.pdf (accessed February 16, 2012).

Breytenbach, J. H. 2004. *Salmonella* control in poultry. *Intervet International b.v.*: 1-4 http://www.safe-poultry.com/documents/ControlofSalmonellainPoultry-August2004.pdf (accessed February 5, 2010).

Bruun, T., G. Sørensen, L. P. Forshell et al. 2009. An outbreak of *Salmonella* Typhimurium infections in Denmark, Norway and Sweden, 2008. *Eurosurveill.*14: pii = 19147.

Buchholz, U., B. Brodhun, S. O. Brockmann et al. 2005. An outbreak of *Salmonella* München in Germany associated with raw pork meat. *J. Food Prot.* 68: 273–276.

Byrd, J. A., R. C. Anderson, T. R. Callaway et al. 2003. Effect of experimental chlorate product administration in the drinking water on *Salmonella* Typhimurium contamination of broilers. *Poultry Sci.* 82: 1403–1406.

Byrd, J. A., B. M. Hargis, D. J. Caldwell et al. 2001. Effect of lactic acid administration in the drinking water during pre-slaughter feed withdrawal on *Salmonella* and *Campylobacter* contamination of broilers. *Poultry Sci.* 80: 278–283.

Callaway, T. R., J. L. Morrow, A. K. Johnson et al. 2005. Environmental prevalence and persistence of *Salmonella* spp. in outdoor swine wallows. *Foodborne Pathog. Dis.* 2: 263–273.

Cardinale, E., S. Maeder, V. Porphyre, and M. Debin. 2010. *Salmonella* in fattening pigs in Reunion Island: Herd prevalence and risk factors for infection. *Prev. Vet. Med.* 96: 281–285.

Cardinale, E., F. Tall, E. F. Guèye, M. Cisse, and G. Salvat. 2004a. Risk factors for *Campylobacter* spp. infection in Senegalese broiler-chicken flocks. *Prev. Vet. Med.* 64: 15–25.

Cardinale, E., F. Tall, E. F. Guèye, M. Cisse, and G. Salvat. 2004b. Risk factors for *Salmonella enterica* subsp. *enterica* infection in senegalese broiler-chicken flocks. *Prev. Vet. Med.* 63: 151–161.

Carli, K. T., A. Eyigor, and V. Caner. 2001. Prevalence of *Salmonella* serovars in chickens in Turkey. *J. Food Prot.* 64: 1832–1835.

CDC. 2009. National *Salmonella* surveillance data—2009. Centers for Disease Control and Prevention. http://www.cdc.gov/ncezid/dfwed/PDFs/SalmonellaAnnualSummaryTables2009.pdf (accessed March 20, 2012).

CDC. 2011. Vital signs: Incidence and trends of infection with pathogens transmitted commonly through food—Foodborne diseases active surveillance network, 10 U.S. sites, 1996–2010. *MMWR Morb. Mortal. Wkly Rep.* 60: 749–755.

Chambers, J. R., J.-R. Bisaillon, Y. Labbé, C. Poppe, and C. F. Langford. 1998. *Salmonella* prevalence in crops of Ontario and Quebec broiler chickens at slaughter. *Poultry Sci.* 77: 1497–1501.

Chaveerach, P., D. A. Keuzenkamp, L. J. A. Lipman, and F. Van Knapen. 2004. Effect of organic acids in drinking water for young broilers on *Campylobacter* infection, volatile fatty acid production, gut microflora and histological cell changes. *Poultry Sci.* 83: 330–334.

Cherrington, C. A., M. Hinton, G. C. Mead, and I. Chopra. 1991. Organic acids: Chemistry, antibacterial activity and practical applications. *Adv. Microb. Physiol.* 32: 87–108.

Christensen, J., D. L. Baggesen, B. Nielsen, and H. Stryhn. 2002. Herd prevalence of *Salmonella* spp. in Danish pig herds after implementation of the Danish *Salmonella* Control Program with reference to a pre-implementation study. *Vet. Microbiol.* 88: 175–188.

Collard, J. M., S. Bertrand, K. Dierick et al. 2007. Drastic decrease of *Salmonella* Enteritidis isolated from humans in Belgium in 2005, shift in phage types and influence on foodborne outbreaks. *Epidemiol. Infect.* 136: 771–781.

Collins, J. D. and P. G. Wall. 2004. Food safety and animal production systems: Controlling zoonoses at farm level. *Sci. Tech. Rev.* 23: 685–700.

Corry, J. E. L., V. M. Allen, W. R. Hudson, and R. H. Davies. 2002. Sources of *Salmonella* on broiler carcasses during transportation and processing: Modes of contamination and methods of control. *J. Appl. Microbiol.* 92: 24–432.

Cox, N. A., J. S. Bailey, L. C. Blackenship, and R. P. Gildersleeve. 1992. Research note: *In ovo* administration of a competitive exclusion culture treatment to broiler embryos. *Poultry Sci.* 71: 1781–1784.

Craven, S. E., N. J. Stern, E. Line, J. S. Bailey, N. A. Cox, and P. Fedorka-Cray. 2000. Determination of the incidence of *Salmonella* spp., *Campylobacter jejuni*, and *Clostridium perfringens* in wild birds near broiler chicken houses by sampling intestinal droppings. *Avian Dis.* 44: 715–720.

Creus, E., J. F. Pérez, B. Peralta, F. Baucells, and E. Mateu. 2007. Effect of acidified feed on the prevalence of *Salmonella* in market-age pigs. *Zoonoses Public Health* 54: 314–319.

Dahl, J. 2008. Feed related interventions in pig herds with a high *Salmonella* sero-prevalence—the Danish experience. *Pig Journal* 61: 6–11.

Davies, P. R., W. E. Morrow, F. T. Jones, J. Deen, P. J. Fedorka-Cray, and J. T. Gray. 1997. Risk of shedding *Salmonella* organisms by market-age hogs in a barn with open-flush gutters. *J. Am. Vet. Assoc.* 210: 386–389.

Davies, R. H. 2005. Pathogen populations on poultry farms. In *Food Safety Control in the Poultry Industry* ed. G. C. Mead, 101–151. Cambridge: Woodhead Publishing Limited.

Davies, R. H., P. J. Health, S. M. Coxon, and A. R. Sayers. 2003. Evaluation of the use of pooled serum, pooled muscle tissue fluid (meat juice) and pooled faeces for monitoring pig herds for *Salmonella*. *J. Appl. Microbiol.* 95: 1016–1025.

Davies, R. H. and M. H. Hinton. 2000. *Salmonella* in animal feed. In *Salmonella in Domestic Animals*, ed. C. Wray and A. Wray, 285–300. Oxford: CAB International.

Davies, R. H., R. Dalziel, J. C. Gibbens et al. 2004. National survey for *Salmonella* in pigs, cattle and sheep at slaughter in Great Britain (1999–2000). *J. Appl. Microbiol.* 95: 750–760.

Delhalle, L., L. De Sadeleer, K. Bollaerts et al. 2008. Risk factors for *Salmonella* and hygiene indicators in the 10 largest Belgian pig slaughterhouses. *J. Food Prot.* 71: 1320–1329.

Directive 2003/99/EC of the European Parliament and of the Council of 17 November 2003 on the monitoring of zoonoses and zoonotic agents, amending Council Decision 90/424/EEC and repealing Council Directive 92/117/EEC. http://eur-lex.europa.eu/LexUriServ/LexUriServ.do?uri=CELEX:32003L0099:EN:NOT (accessed February 15, 2012).

Dorn-in, S., R. Fries, P. Padungtod et al. 2009. A cross-sectional study of *Salmonella* in pre-slaughter pigs in a production compartment of northern Thailand. *Prev. Vet. Med.* 88: 15–23.

Doyle, M. P. and M. C. Erickson. 2006. Reducing the carriage of foodborne pathogens in livestock and poultry. *Poultry Sci.* 85: 960–973.

Dunkley, K. D., T. R. Callaway, V. I. Chalova et al. 2009. Foodborne *Salmonella* ecology in the avian gastrointestinal tract. *Anaerobe* 15: 26–35.

EC. 2001. Commission decision of 8 June 2001. *Off. J. Eur. Comm.* L 165: 48–53.

EFSA. 2004. Opinion of the Scientific panel on biological hazards on the requests from the Commission related to the use of vaccines for the control of Salmonella in poultry. *EFSA Journal* 114: 1–74. http://www.efsa.europa.eu/en/scdocs/doc/opinion_biohaz15_ej114_vacc_salminpoultry_v2_en1,2.pdf (accessed February 3, 2010).

EFSA. 2006. Opinion of the Scientific Panel on Biological Hazards on the request from the Commission related to "Risk assessment and mitigation options of *Salmonella* in pig production." *EFSA Journal* 341: 1–131.

EFSA. 2007. The community summary report on trends and sources of zoonoses, zoonotic agents, antimicrobial resistance and foodborne outbreaks in the European Union in 2006. *EFSA Journal* 130: 1–352.

EFSA. 2008. Microbiological risk assessment in feeding stuffs for food-producing animals: Scientific opinion of the panel on biological hazards. *EFSA Journal* 720: 1–84.

EFSA. 2009. Analysis of the baseline survey on the prevalence of *Salmonella* in holdings with breeding pigs in the EU, 2008. Part A: *Salmonella* prevalence estimates. *EFSA Journal* 7: 1–93.

EFSA. 2010. Scientific opinion on a quantitative microbiological risk assessment of *Salmonella* in slaughter and breeder pigs. *EFSA Journal* 8: 1547.

EFSA Denmark. 2006. Trends and sources of zoonoses and zoonotic agents in humans, foodstuffs, animals and feedingstuffs including information on foodborne outbreaks and antimicrobial. *EFSA Journal* 12: 8.

EFSA Finland. 2006. Trends and sources of zoonoses and zoonotic agents in humans, foodstuffs, animals and feedingstuffs including information on foodborne outbreaks and antimicrobial. *EFSA Journal* 12: 6.

EFSA Sweden. 2006. Trends and sources of zoonoses and zoonotic agents in humans, foodstuffs, animals and feedingstuffs including information on foodborne outbreaks and antimicrobial. *EFSA Journal* 12: 8.

Ellis-Iversen, J., F. Jorgensen, S. Bull, L. Powell, A. J. Cook, and T. J. Humphrey. 2009. Risk factors for *Campylobacter* colonization during rearing of broiler flocks in Great Britain. *Prev. Vet. Med.* 89: 178–184.

Escudero-Gilete, M. L., M. L. González-Miret, R. M. Temprano, and F. J. Hereda. 2007. Application of a multivariate concentric method system for the location of *Listeria monocytogenes* in a poultry slaughterhouse. *Food Control* 18: 69–75.

EU. 2003. Regulation (EC) No 2160/2003 of the European parliament and of the council of 17 November 2003 on the control of salmonella and other specified foodborne zoonotic agents. *Off. J. Eur. Union* L 325: 1–15.

Evans, S. J. and A. R. Sayers. 2000. A longitudinal study of *Campylobacter* infection of broiler flocks in Great Britain. *Prev. Vet. Med.* 46: 209–223.

FAO. 2008. Biosecurity for highly pathogenic avian influenza. FAO animal production and health paper No. 165. ftp://ftp.fao.org/docrep/fao/011/i0359e/i0359e00.pdf (accessed March 16, 2011).

FAO. 2010. Agricultural production. FAO statistical yearbook. http://www.fao.org/economic/ess/ess-publications/ess-yearbook/ess-yearbook2010/yearbook2010-production/en/ (accessed March 16, 2010).

Farzan, A., R. M. Friendship, C. E. Dewey, K. Warriner, C. Poppe, and K. Klotins. 2006. Prevalence of *Salmonella* spp. on Canadian pig farms using liquid or dry-feeding. *Prev. Vet. Med.* 73: 241–254.

Feder, I., J. C. Nietfeld, J. Galland et al. 2001. Comparison of cultivation and PCR-hybridization for detection of *Salmonella* in porcine fecal and water samples. *J. Clin. Microbiol.* 39: 2477–2484.

Fedorka-Cray, P. J., J. T. Gray, and C. Wray. 2000. Salmonella infections in pigs. In *Salmonella in Domestic Animals,* ed. C. Wray and A. Wray, 191–207. Oxford: CAB International.

Fiala, N. 2008. Meeting the demand: An estimation of potential future greenhouse gas emissions from meat production. *Ecol. Econ.* 67: 412–419.

Foley, S. L. and A. M. Lynne. 2008. Food animal-associated *Salmonella* challenges: Pathogenicity and antimicrobial resistance. *J. Anim. Sci.* 86: E173–E187.

FSIS. 1996. Federal Register: Part 11. Pathogen reduction: hazard analysis and critical control point (HACCP) system: Final rule, 38806-38989. http://www.fsis.usda.gov/OPPDE/rdad/FRPubs/93-016F.htm (accessed February 21, 2012).

FSIS. 2011. Quarterly progress report on Salmonella testing of selected raw meat and poultry products: Preliminary results, January–March 2011. http://www.fsis.usda.gov/PDF/Q1_2011_Salmonella_Testing.pdf#page=1 (accessed March 20, 2012).

Funk, J. and W. A. Gebreyes. 2004. Risk factors associated with *Salmonella* prevalence on swine farms. *J. Swine Health Manage.* 12: 246–251.

García-Feliz, C., J. A. Collazos, A. Carvajal et al. 2007. *Salmonella enterica* infections in Spanish swine fattening units. *Zoonoses Public Health* 54: 294–300.

García-Feliz, C., A. Carvajal, J. A. Collazos, and P. Rubio. 2009. Herd-level risk factors for faecal shedding of *Salmonella enterica* in Spanish fattening pigs. *Prev. Vet. Med.* 91: 130–136.

Gebreyes, W. A., S. Thakur, P. R. Davies, J. A. Funk, and C. Altier. 2004. Trends in antimicrobial resistance, phage types and integrons among *Salmonella* serotypes from pigs, 1997–2000. *J. Antimicrob. Chemother.* 53: 997–1003.

Gibbens, J. C., S. J. S. Pascoe, S. J. Evans, R. H. Davies, and A. R. Sayers. 2001. A trial of biosecurity as a means to control *Campylobacter* infection of broiler chickens. *Prev. Vet. Med.* 48: 85–99.

Gibson, H., J. H. Taylor, K. E. Hall, and J. T. Holah. 1999. Effectiveness of cleaning techniques used in the food industry in terms of the removal of bacterial biofilms. *J. Appl. Microbiol.* 87: 41–48.

Gómez-Laguna, J., M. Hernández, E. Creus et al. 2011. Prevalence and antimicrobial susceptibility of *Salmonella* infections in free-range pigs. *Vet. J.* 190: 176–178.

Gradel, K. O. and E. Rattenborg. 2003. A questionnaire-based, retrospective field study of persistence of *Salmonella* Enteritidis and *Salmonella* Typhimurium in Danish broiler houses. *Prev. Vet. Med.* 56: 267–284.

Graham, J. P., J. H. Leibler, L. B. Price et al. 2008. The animal-human interface and infectious disease in industrial food animal production: Rethinking biosecurity and biocontainment. *Public Health Rep.* 123: 282–299.

Guo, C., R. M. Hoekstra, C. M. Schroeder et al. 2011. Application of Bayesian techniques to model the burden of human salmonellosis attributable to U.S. food commodities at the point of processing: Adaptation of a Danish model. *Foodborne Pathog. Dis.* 8: 509–516.

Hald, B., A. Olsen, and M. Madsen. 1998. *Typhaea stercorea* (Cleoptera: Mycetophagidae), a carrier of *Salmonella enterica* serovar Infantis in a Danish broiler house. *J. Econ. Entomol.* 91: 660–664.

Hald, B., E. Rattenborg, and M. Madsen. 2001. Role of depletion of broiler houses on the occurrence of *Campylobacter* spp. in chicken flocks. *Lett. Appl. Microbiol.* 32: 253–256.

Hald, B., H. Skovgård, D. D. Bang et al. 2004. Flies and *Campylobacter* infection of broiler flocks. *Emerg. Infect. Dis.* 10: 1490–1492.

Hald, T. 2008. EU-wide baseline studies: achievements and difficulties faced. *Trends Food Sci. Technol.* 19: S40–S48.

Hänninen, M.-L., P. Perko-Mäkelä, A. Pitkälä, and H. Rautelin. 2000. A three-year study of *Campylobacter jejuni* genotypes in humans with domestically acquired infections and in chicken samples from the Helsinki area. *J. Clin. Microbiol.* 38: 1998–2000.

Havelaar, A. H., M.-J. J. Mangen, A. A. de Koeijer et al. 2007. Effectiveness and efficiency of controlling *Campylobacter* on broiler chicken meat. *Risk Anal.* 27: 831–844.

Hedemann, M. S., L. L. Mikkelsen, P. J. Naughton, and B. B. Jensen. 2005. Effect of feed particle size and feed processing on morphological characteristics in the small and large intestine of pigs and on adhesion of *Salmonella enterica* serova Typhimurium DT12 in the ileum in vitro. *J. Anim. Sci.* 83: 1554–1562.

Hein, I., C. Schneck, M. Knögler et al. 2003. *Campylobacter jejuni* isolated from poultry and humans in Styria, Austria: Epidemiology and ciprofloxacin resistance. *Epidemiol. Infect.* 130: 377–386.

Hensel, A. and H. Neubauer. 2002. Human pathogens associated with on-farm practices – Implications for control and surveillance strategies. In *Food Safety Assurance in the Pre-Harvest Phase*, ed. F. J. M. Smulders and J. D. Collins, 125–139, Vol. 1. Wageningen: Wageningen Academic Publishers.

Hernandez, J., J. Bonnedahl, J. Waldenström, H. Palmgren, and B. Olsen 2003. *Salmonella* in birds migrating through Sweden. *Emerg. Infect. Dis.* 9: 753–755.

Hofshagen, M. and H. Kruse. 2005. Reduction in flock prevalence of *Campylobacter* spp. in broilers in Norway after implementation of an action plan. *J. Food Prot.* 68: 2220–2223.

Hoover, N. J., P. B. Kenney, J. D. Amick, and W. A. Hypes. 1997. Preharvest sources of *Salmonella* colonization in turkey production. *Poultry Sci.* 76: 1232–1238.

Huang, D. S., D. F. Li, J. J. Xing, Y. X. Ma, Z. J. Li, and S. Q. Lv. 2006. Effects of feed particle size and feed form on survival of *Salmonella typhimurium* in the alimentary tract and cecal *S. typhimurium* reduction in growing broilers. *Poultry Sci.* 87: 405–420.

Humphrey, T. J., A. Henley, and D. G. Lanning. 1993. The colonization of broiler chickens with *Campylobacter jejuni*: Some epidemiological investigations. *Epidemiol. Infect.* 110: 601–607.

Humphrey, T., S. O'Brien, and M. Madsen. 2007. *Campylobacters* as zoonotic pathogens: A food production perspective. *Int. J. Food Microbiol.* 117: 237–257.

Huneau-Salaün, A., M. Chemaly, S. Le Bouquin et al. 2009. Risk factors for *Salmonella enterica* subsp. *enterica* contamination in 519 French laying hen flocks at the end of the laying period. *Prev. Vet. Med.* 89: 51–58.

Huneau-Salaün, A., M. Denis, L. Balaine, and G. Salvat. 2007. Risk factors for *Campylobacter* spp. colonization in French free-range broiler-chicken flocks at the end of the indoor rearing period. *Prev. Vet. Med.* 80: 34–48.

Hurd, H. S., J. D. McKean, I. V. Wesley, and L. A. Karriker. 2001. The effect of lairage on *Salmonella* isolation from market swine. *J. Food Prot.* 64: 939–944.

Hurd, H. S., C. Enøe, L. Sørenson et al. 2008. Risk-based analysis of the Danish pork *Salmonella* program: Past and future. *Risk Anal.* 28: 341–351.

Huttunen, A., T. Johansson, P. Kostamo et al. 2006. *Salmonella Control and Occurrence of Salmonella from 1995 to 2004 in Finland.* Finnish Food Safety Authority Evira, Helsinki, 1–95.

Jacobs-Reitsma, W. F., N. M. Bolder, and R. W. Mulder. 1994. Cecal carriage of *Campylobacter* and *Salmonella* in Dutch broiler flocks at slaughter: A one-year study. *Poultry Sci.* 73: 1260–1266.

Jacobs-Reitsma, W. 2000. *Campylobacter* in the food supply. In *Campylobacter,* ed. I. Nachamkin and M. J. Blaser, 467–481, 2nd ed. Washington, D.C.: American Society for Microbiology Press.

Jansen, A., C. Frank, and K. Stark. 2007. Pork and pork products as a source for human salmonellosis in Germany. *Berl Munch Tierarztl Wochenschr.* 120: 340–346.

Jeffrey, J. S., E. R. Atwill, and A. Hunter, A. 2001. Farm and management variables linked to fecal shedding of *Campylobacter* and *Salmonella* in commercial squab production. *Poultry Sci.* 80: 66–70.

Jensen, A. N., J. Lodal, and D. L. Baggesen. 2004. High diversity of *Salmonella* serotypes found in an experiment with outdoor pigs. *NJAS—Wageningen J. Life Sci.* 52: 109–117.

Jiménez, M., P. Soler, J. D. Venanzil, P. Canté, C. Varela, and F. Martínez-Navarro. 2005. An outbreak of *Campylobacterjejuni* enteritis in a school of Madrid, Spain. *Eurosurveill.*10: pii = 533.

Johnsen, G., H. Kruse, and M. Hofshagen. 2006. Genetic diversity and description of transmission routes for *Campylobacter* on broiler farms by amplified-fragment length polymorphism. *J. Appl. Microbiol.* 101: 1130–1139.

Jones, F. T. and K. E. Richardson. 2004. *Salmonella* in commercially manufactured feeds. *Poultry Sci.* 83: 384–391.

Jore, S., T. M. Lyngstad, M. Hofshagen et al. 2009. The surveillance and control programme for *Salmonella* in live animals, eggs and meat in Norway. In E. Brun, H. Hellbergm, and S. Sviland. (eds.), 1–10. Oslo: National Veterinary Institute, Oslo.

Jung, Y. S., R. C. Anderson, J. A. Byrd et al. 2003. Reduction of *Salmonella* Typhimurium in experimentally challenged broilers by nitrate adaptation and chlorate supplementation in drinking water. *J. Food Prot.* 66: 660–663.

Kapperud, G., E. Skjerve, L. Vik et al. 1993. Epidemiological investigation of risk factors for *Campylobacter* colonization in Norwegian broiler flocks. *Epidemiol. Infect.* 111: 245–255.

Katsma, W. E. A., A. A. De Koeijer, W. F. Jacobs-Reitsma, M.-J. J. Mangen, and J. A. Wagenaar. 2007. Assessing interventions to reduce the risk of *Campylobacter* prevalence in broilers. *Risk Anal.* 27: 863–876.

Kich, J. D., A. Coldebella, N. Morés et al. 2011. Prevalence, distribution, and molecular characterization of *Salmonella* recovered from swine finishing herds and a slaughter facility in Santa Catarina, Brazil. *Int. J. Food Microbiol.* 151: 307–313.

Kim, A., Y. J. Lee, M. S. Kang, S. I. Kwag, and J. K. Cho. 2007. Dissemination and tracking of *Salmonella* spp. in integrated broiler operation. *J. Vet. Sci.* 8: 155–161.

Lahellac, C., P. Colin, G. Bennejean, J. Paquin, A. Guillerm, and J. C. Debois. 1986. Influence of resident *Salmonella* on contamination of broiler flocks. *Poultry Sci.* 65: 2034–2039.

Lázaro, N. S., A. Tibana, and E. Hofer. 1997. *Salmonella* spp. in healthy swine and in abattoir environments in Brazil. *J. Food Prot.* 60: 1029–1033.

Lewerin, S. S., Boqvist, B. Engström, and P. Häggblom. 2005. The effective control of *Salmonella* in Swedish poultry. In *Food Safety Control in the Poultry Industry*, ed. G. C. Mead, 195–215. Cambridge: Woodhead Publishing Limited.

Letellier, A., G. Beauchamp, E. Guévremont, S. D'allaire, D. Hurnik, and S. Quessy. 2009. Risk factors at slaughter associated with presence of *Salmonella* on hog carcasses in Canada. *J. Food Prot.* 72: 2326–2331.

Lindblom, G.-B., E. Sjögren, and B. Kaijser. 1986. Natural campylobacter colonization in chickens raised under different environmental conditions. *J. Hyg.* 96: 385–391.

Lo Fo Wong, D. M. A., T. Hald, P. J. van der Wolf, and M. Swanenburg. 2002. Epidemiology and control measures for *Salmonella* in pigs and pork. *Livest. Prod. Sci.* 76: 215–222.

Lo Fo Wong, D. M. A., J. Dahl, H. Stege et al. 2004. Herd-level risk factors for subclinical *Salmonella* infection in European finishing-pig herds. *Prev. Vet. Med.* 62: 253–266.

Madec, F., F. Humbert, G. Salvat, and P. Maris. 1999. Measurement of the residual contamination of post-weaning facilities for pigs and related risk factors. *J. Vet. Med.* 46: 47–56.

Mainar-Jaime, R. C., N. Atashparvar, M. Chirino-Trejo, and K. Rahn. 2008. Survey on *Salmonella* prevalence in slaughter pigs from Saskatchewan. *Can. Vet. J.* 49: 793–796.

Majowicz, S. E., J. Musto, E. Scallan et al. 2010. The global burden of nontyphoidal *Salmonella* gastroenteritis. *Clin. Infect. Dis.* 50: 882–889.

Mannion, C., J. Fanning, J. McLernon et al. 2012. The role of transport, lairage and slaughter processes in the dissemination of *Salmonella* spp. in pigs in Ireland. *Food Res. Int.* 45: 871–879.

Marin, C., A. Hernandiz, and M. Lainez. 2009. Biofilm development capacity of *Salmonella* strains isolated in poultry risk factors and their resistance against disinfectants. *Poultry Sci.* 88: 424–431.

Mastroeni, P., J. A. Chabalgoity, S. J. Dunstan, D. J. Maskell, and G. Dougan. 2001. *Salmonella*: Immune responses and vaccines. *Vet. J.* 161: 132–164.

Maurer, A. J. 2003. Chicken. In *Encyclopedia of Food Sciences and Nutrition*, ed. B. Caballero, L. Trugo, and P. M. Finglas, 4680–4686. Amsterdam, Elsevier Science Ltd.

Mazick, A., S. Ethelberg, E. M. Nielsen, K. Molbak, and M. Lisby. 2006. An outbreak of *Campylobacter jejuni* associated with consumption of chicken, Copenhagen, 2005. *Eurosurveill.*11: pii = 622. http://www.eurosurveillance.org/ViewArticle. aspx?ArticleId=622 (accessed December 30, 2009).

McCapes, R. H., H. E. Ekperigin, W. J. Cameron et al. 1989. Effect of a new pelleting process on the level of contamination of poultry mash by *Escherichia coli* and *Salmonella*. *Avian Dis.* 33: 103–111.

McDowell, S. W. J., F. D. Menzies, S. H. McBride et al. 2008. *Campylobacter* spp. in conventional broiler flocks in Northern Ireland: Epidemiology and risk factors. *Prev. Vet. Med.* 84: 261–276.

Mead, G. C. 2000. Review: Prospects for "competitive exclusion" treatment to control *Salmonellas* and other foodborne pathogens in poultry. *Vet. J.* 159: 111–123.

Mead, P., L. Slutsker, V. Dietz et al. 1999. Food-related illness and death in the United States. *Emerg. Infect. Dis.* 5: 607–625.

Meremäe, K., P. Elias, T. Tamme et al. 2010. The occurrence of *Campylobacter* spp. in Estonian broiler chicken production in 2002–2007. *Food Control* 21: 272–275.

Messens, W., L. Herman, L. De Zutter, and M. Heyndrickx. 2009. Multiple typing for the epidemiological study of contamination of broilers with thermotolerant *Campylobacter. Vet. Microbiol.* 138: 120–131.

Mikkelsen, L. L., P. J. Naughton, M. S. Hedemann, and B. B. Jensen. 2004. Effects of physical properties of feed on microbial ecology and survival of *Salmonella enterica* serovar Typhimurium in the pig gastrointestinal tract. *Appl. Environ. Microbiol.* 70: 3485–3492.

Missotten, J. A. M., J. Goris, J. Michiels et al. 2009. Screening of isolated lactic acid bacteria as potential beneficial strains for fermented liquid pig feed production. *Anim. Feed Sci. Technol.* 150: 122–138.

Molla, B., A. Sterman, J. Mathews et al. 2010. *Salmonella enterica* in commercial swine feed and subsequent isolation of phenotypically and genotypically related strains from fecal samples. *Appl. Environ. Microbiol.* 76: 7188–7193.

Mollenhorst, H., C. J. van Woudenbergh, E. G. M. Bokkers, and I. J. M. de Boer. 2005. Risk factors for *Salmonella enteritidis* infections in laying hens. *Poultry Sci.* 84: 1308–1313.

Murase, T., K. Senjyu, T. Maeda et al. 2001. Monitoring of chicken houses and an attached egg-processing facility in a laying farm for *Salmonella* contamination between 1994 and 1998. *J. Food Prot.* 64: 1912–1916.

Nakyinsige, K., Y. Che Man, and A. Q. Sazili.(2012). Halal authenticity issues in meat and meat products. *Meat Sci.*

Namata, H., E. Méroc, M. Aerts et al. 2008. *Salmonella* in Belgian laying hens: An identification of risk factors. *Prev. Vet. Med.* 83: 323–336.

Namata, H., S. Welby, M. Aerts et al. 2009. Identification of risk factors for the prevalence and persistence of *Salmonella* in Belgian broiler chicken flocks. *Prev. Vet. Med.* 90: 211–222.

Newell, D. G. and C. Fearnley. 2003. Sources of *Campylobacter* colonization in broiler chickens. *Appl. Environ. Microbiol.* 69: 4343–4351.

Nollet, N., D. Maes, L. De Zutter et al. 2004. Risk factors for the herd-level bacteriologic prevalence of *Salmonella* in Belgian slaughter pigs. *Prev. Vet. Med.* 65: 63–75.

Nylen, G., F. Dunstan, S. R. Palmer et al. 2002. The seasonal distribution of *Campylobacter* infection in nine European countries and New Zealand. *Epidemiol. Infect.* 128: 383–390.

Ojha, S. and M. Kostrzynska. 2007. Approaches for reducing *Salmonella* in pork production. *J. Food Prot.* 70: 2676–2694.

Österberg, J., I. Vågsholm, S. Boqvist, and S. Sternberg Lewerin. 2006. Feed-borne outbreak of *Salmonella* Cubana in Swedish pig farms: Risk factors and factors affecting the restriction period in infected farms. *Acta Vet. Scand.* 47: 13–22.

Pearce, R. A., D. J. Bolton, J. J. Sheridan, D. A. McDowell, I. S. Blair, and D. Harrington. 2004. Studies to determine the critical control points in pork slaughter hazard analysis and critical control point systems. *Int. J. Food Microbiol.* 90: 331–339.

Pearson, A. D., M. H. Greenwood, J. Donaldson et al. 2000. Continuous source outbreak of campylobacteriosis traced to chicken. *J. Food Prot.* 63: 309–314.

Pearson, A. D., M. Greenwood, T. D. Healing et al. 1993. Colonization of broiler chickens by waterborne *Campylobacter jejuni*. *Appl. Environ. Microbiol.* 59: 987–996.

Pearson, A. D., M. H. Greenwood, and R. K. A. Feltham. 1996. Microbial ecology of *Campylobacter jejuni* in a United Kingdom chicken supply chain: Intermittent common source, vertical transmission and amplification by flock propagation. *Appl. Environ. Microbiol.* 62: 4614–4620.

Perko-Mäkela, P., M. Hakkinen, T. Honkanen-Buzalski, and M.-L. Hänninen. 2002. Prevalence of *Campylobacters* in chicken flocks during the summer of 1999 in Finland. *Epidemiol. Infect.* 129: 187–192.

Petersen, L. and A. Wedderkopp. 2001. Evidence that certain clones of *Campylobacter jejuni* persist during successive broiler flock rotations. *Appl. Environ. Microbiol.* 67: 2739–2745.

Phan, T. T., L. T. L. Khai, N. Ogasawara et al. 2005. Contamination of *Salmonella* in retail meats and shrimps in the Mekong Delta, Vietnam. *J. Food Prot.* 65: 1077–1080.

Pinto, C. J. and V. S. Urcelay. 2003. Biosecurity practices on intensive pig production systems in Chile. *Prev. Vet. Med.* 59: 139–145.

Poljak, Z., C. E. Dewey, R. M. Friendship, S. W. Martin, and J. Christensen. 2008. Multilevel analysis of risk factors for *Salmonella* shedding in Ontario finishing pigs. *Epidemiol. Infect.* 136: 1388–1400.

Prohászka, L., B. M. Jayarao, A. Fábián, and S. Kovács. 1990. The role of intestinal volatile fatty acids in the *Salmonella* shedding of pigs. *J. Vet. Med.* 37: 570–574.

Proux, K., R. Cariolet, P. Fravalo, C. Houdayer, A. Keranflech, and F. Madec. 2001. Contamination of pigs by nose-to-nose contact or airborne transmission of *Salmonella* Typhimurium. *Vet. Res.* 32: 591–600.

Racicot, M., D. Venne, A. Durivage, and J.-P.Vaillancourt. 2011. Description of 44 biosecurity errors while entering and exiting poultry barns based on video surveillance in Quebec, Canada. *Prev. Vet. Med.* 100: 193–199.

Rajić, A., J. Keenliside, M. E. McFall et al. 2005. Longitudinal study of *Salmonella* species in 90 Alberta swine finishing farms. *Vet. Microbiol.* 105: 47–56.

Rathgeber, B. M., K. L. Thompson, C. M. Ronalds, and K. L. Budgell. 2009. Microbiological evaluation of poultry house wall materials and industrial cleaning agents. *J. Appl. Poult. Res.* 18: 579–582.

Refrégier-Petton, J., N. Rose, M. Denis, and G. Salvat. 2001. Risk factors for *Campylobacter* spp. in contamination in French broiler-chicken flocks at the end of the rearing period. *Prev. Vet. Med.* 50: 89–100.

Regulation (EC) No. 183/2005 of the European Parliament and of the Council of 12 January 2005 laying down requirements for feed hygiene. *Off. J. Eur. Union* L 35/1. http://eur-lex.europa.eu/LexUriServ/LexUriServ.do?uri = OJ:L:2005:035:0001:0022:EN:PDF (accessed February 22, 2012).

Regulation (EC) No. 853/2004 of the European Parliament and of the Council of April 29, 2004 laying down specific hygiene rules for food of animal origin. *Off. J. Eur. Union* L 226/22. http://www.fsai.ie/uploadedFiles/Reg853_2004(1).pdf (accessed February 22, 2012).

Regulation (EC) No. 2160/2003 of the European Parliament and of the Council of 17 November 2003 on the control of *Salmonella* and other specified food-borne zoonotic agents. *Official Journal of the European Union* L 325/1. http://eur-lex.europa.eu/LexUriServ/LexUriServ.do?uri = OJ:L:2003:325:0001:0015:EN:PDF (accessed February 15, 2012).

Renwick, S. A., R. J. Irwin, R. C. Clarke, W. B. McNab, C. Poppe, and S. A. McEwen. 1992. Epidemiological associations between characteristics of registered broiler chicken flocks in Canada and the *Salmonella* culture status of floor litter and drinking water. *Can. Vet. J.* 33: 449–458.

Rho, M.-J., M.-S. Chung, J.-H. Lee, and J. Park. 2001. Monitoring of microbial hazards at farms, slaughterhouses, and processing lines of swine in Korea. *J. Food Prot.* 64: 1388–1391.

Ricke, S. C. 2003. Perspectives on the use of organic acids and short chain fatty acids as antimicrobials. *Poultry Sci.* 82: 632–639.

Ricke, S. C. 2005. Ensuring the safety of poultry feed. In *Food Safety Control in the Poultry Industry*, ed. G. C. Mead, 174–194. Cambridge: Woodhead Publishing Limited.

Roche, A. J., N. A. Cox, L. Richardson et al. 2009. Transmission of *Salmonella* to broilers by contaminated larval and adult lesser mealworms, *Alphitobius diaperinus* (Coleoptera: Tenebrionidae). *Poultry Sci.* 88: 44–48.

Rocourt, J., G. Moy, K. Vierk, and J. Schlundt. 2003. The present state of foodborne disease in OECD countries. Food Safety Department, World Health Organization: Geneva. http://www.who.int/foodsafety/publications/foodborne_disease/oecd_fbd.pdf (accessed December 31, 2009).

Roesler, U., A. Von Altrock, P. Heller et al. 2005. Effects of fluorequinolone treatment acidified feed, and improved hygiene measures on the occurrence of *Salmonella* Typhimurium DT104 in an integrated pig breeding herd. *J. Vet. Med.* 52: 69–74.

Rostagno, M. H. 2009. Can stress in farm animals increase food safety risk? *Foodborne Pathog. Dis.* 6: 767–776.

Rostagno, M. H. and T. R. Callaway. 2012. Pre-harvest risk factors for *Salmonella* enterica in pork production. *Food Res. Int.* 45: 634–640.

Santos, F. B. O., B. W. Sheldon, A. A. Santos, Jr., and P. R. Ferket. 2008. Influence of housing system, grain type, and particle size on *Salmonella* colonization and shedding of broilers fed triticale or corn-soybean meal diets. *Poultry Sci.* 87: 405–420.

Scallan, E., R. M. Hoekstra, F. J. Angulo et al. 2011. Foodborne illness acquired in the United States—Major pathogens. *Emerg. Infect. Dis.* 17: 7–15.

Schmidt, P. L., A. M. O'Connor, J. D. McKean, and H. S. Hurd. 2004. The association between cleaning and disinfection of lairage pens and the prevalence of *Salmonella enterica* in swine at harvest. *J. Food Prot.* 67: 1384–1388.

Shirota, K., H. Katoh, T. Murase, T. Ito, and K. Otsuki. 2001. Monitoring of layer feed and eggs for *Salmonella* in Eastern Japan between 1993 and 1998. *J. Food Prot.* 64: 734–737.

Shreeve, J. E., M.,, Toszeghy, A. Ridley, and D. G. Newell. 2002. The carry-over of *Campylobacter* isolates between sequential poultry flocks. *Avian Dis.* 46: 378–385.

Sirsat, S. A., A. Muthaiyan, and S. C. Ricke. 2009. Antimicrobials for foodborne pathogen reduction in organic and natural poultry production. *J. Appl. Poult. Res.* 18: 379–388.

Slader, J., G. Domingue, F. Jørgensen, K. McAlpine, R. J. Owen, F. J. Bolton, and T. J. Humphrey. 2002. Impact of transport crate reuse and of catching and processing on *Campylobacter* and *Salmonella* contamination of broiler chickens. *Appl. Environ. Microbiol.* 68: 713–719.

Small, A., C.-A. Reid, and S. Buncic. 2003. Conditions in lairages at abbattoirs for ruminants in southwest England and in vitro survival of *Escherichia coli* O157, *Salmonella* Kedougou and *Campylobacter jejuni* on lairage-related substrates. *J. Food Prot.* 66: 1570–1575.

Snow, L. C., R. H. Davies, K. H. Christiansen et al. 2008. Survey of the prevalence of *Salmonella* on commercial broiler farms in the United Kingdom, 2005/6. *Vet. Rec.* 163: 649–654.

Snow, L. C., R. H. Davies, K. H. Christiansen et al. 2007. Survey of the prevalence of *Salmonella* species on commercial laying farms in the United Kingdom. *Vet. Rec.* 161: 471–476.

Spoolder, H. A. M., S. A. Edwards, and S. Corning. 2000. Legislative methods for specifying stocking density and consequences for the welfare of finishing pigs. *Livest. Prod. Sci.* 64: 167–173.

Stern, N. J., N. A. Cox, and M. T. Musgrove. 2001. Incidence and levels of *Campylobacter* in broilers after exposure to an inoculated seeder bird. *J. Appl. Poult. Res.* 10: 315–318.

Stern, N. J., M. C. Robach, N. A. Cox, and M. T. Musgrove. 2002. Effect of drinking water chlorination on *Campylobacter* spp. colonization of broilers. *Avian Dis.* 46: 401–404.

Stewart, P. S., J. Rayner, F. Roe, and W. M. Rees. 2001. Biofilm penetration and disinfection efficacy of alkaline hypochlorite and chlorosulfamates. *J. Appl. Microbiol.* 91: 525–532.

Studer, E., J. Lüthy, and P. Hübner. 1999. Study of the presence of *Campylobacter jejuni* and *C. coli* in sand samples from four Swiss chicken farms. *Res. Microbiol.* 150: 213–219.

Suzuki, H. and S. Yamamoto. 2009. *Campylobacter* contamination in retail poultry meats and by-products in Japan: A literature survey. *Food Control* 20: 531–537.

Swanenburg, M., H. A. P. Urlings, D. A. Keuzenkamp, and J. M. A. Snijders. 2001a. *Salmonella* in the lairage of pig slaughterhouses. *J. Food Prot.* 64: 12–16.

Swanenburg, M., H. A. P. Urlings, J. M. A. Snijders, D. A. Keuzenkamp, and F. van Knapen 2001b. *Salmonella* in slaughter pigs: Prevalence, serotypes and critical control points during slaughter in two slaughterhouses. *Int. J. Food Microbiol.* 70: 243–254.

Tavechio, A. T., A. C. R. Ghilardi, J. T. M. Peresi et al. 2002. *Salmonella* serotypes isolated from nonhuman sources in São Paulo, Brazil, from 1996 through 2000. *J. Food Prot.* 65: 1041–1044.

Teirlynck, E., F. Haesebrouck, F. Pasmans, J. Dewulf, R. Ducatelle, and F. Van Immerseel. 2009. The cereal type in feed influences *Salmonella* Enteritidis colonization in broilers. *Poultry Sci.* 88: 2108–2112.

Thompson, J. L. and M. Hinton. 1997. Antibacterial activity of formic and propionic acids in the diet of hens on *Salmonellas* in the crop. *Brit. Poult. Sci.* 38: 59–65.

Tinker, D. B., C. H. Burton, and V. M. Allen. 2005. Catching, transporting and lairage of live poultry. In *Food Safety Control in the Poultry Industry*, ed. G. C. Mead, 153–173. Cambridge: Woodhead Publishing Limited.

Totton, S. C., A. M. Farrar, W. Wilkins et al. 2012. A systematic review and meta-analysis of the effectiveness of biosecurity and vaccination in reducing *Salmonella* spp. in broiler chickens. *Food Res. Int.* 45: 617–627.

Tsola, E., E. H. Drosinos, and P. Zoiopoulos. 2008. Impact of poultry slaughter house modernization and updating of food safety management systems on the microbiological quality and safety of products. *Food Control* 19: 423–431.

Udayamputhoor, R. S., H. Hariharan, T. A. van Lunen et al. 2003. Effects of diet formulations containing proteins from different sources on intestinal colonization by *Campylobacter jejuni* in broiler chickens. *Can. J. Vet. Res.* 67: 204–212.

Valdezate, S., A. Vidal, S. Herrera-León et al. 2005. *Salmonella* Derby clonal spread from pork. *Emerg. Infect. Dis.* 11: 694–698.

Van de Giessen, A., S.-I. Mazurier, W. Jacobs-Reitsma et al. 1992. Study on the epidemiology and control of *Campylobacter jejuni* in poultry broiler flocks. *Appl. Environ. Microbiol.* 58: 1913–1917.

Van de Giessen, A., J. J. H. C. Tilburg, W. S. Ritmeester, and J. Van der Plas. 1998. Reduction of *Campylobacter* infections in broiler flocks by application of hygiene measures. *Epidemiol. Infect.* 121: 57–66.

Van der Gaag, M., F. Vos, H. W. Saatkamp, M. van Boven, P. van Beek, and P. Huirne. 2004. A state-transition simulation model for the spread of *Salmonella* in the pork supply chain. *Eur. J. Oper. Res.* 156: 782–798.

Van der Wolf, P. J., W. B. Wolbers, A. R. W. Elbers et al. 2001a. Herd level husbandry factors associated with the serological *Salmonella* prevalence in finishing pig herds in The Netherlands. *Vet. Microbiol.* 78: 205–219.

Van der Wolf, P. J., F. W. van Schie, A. R. W. Elbers. 2001b. Epidemiology: Administration of acidified drinking water to finishing pigs in order to prevent *Salmonella* infections. *Vet. Q.* 23: 121–125.

Van der Wolf, P. J., A. R. W. Elbers, H. M. J. F. van der Heijden, F. W. van Schie, W. A. Hunneman, and M. J. M. Tielen. 2001c. *Salmonella* seroprevalence at the population and herd level in pigs in The Netherlands. *Vet. Microbiol.* 80: 171–184.

Van Immerseel, F., J. De Buck, F. Pasmans et al. 2004. Intermittent long-term shedding and induction of carrier birds after infection of chickens early posthatch with a low or high dose of *Salmonella* Enteritidis. *Poultry Sci.* 83: 1911–1916.

Van Immerseel, F., L. De Zutter, K. Houf, F. Pasmans, F. Haesebrouck, and R. Ducatelle. 2009. Strategies to control *Salmonella* in the broiler production chain. *Worlds Poult. Sci. J.* 65: 367–392.

Van Immerseel, F., J. B. Russell, M. D. Flythe et al. 2006. The use of organic acids to combat *Salmonella* in poultry: A mechanistic explanation of the efficacy. *Avian Pathol.* 35: 182–188.

Veling, J., H. Wilpshaar, K. Frankena, C. Bartels, and H. W. Barkema. 2002. Risk factors for clinical *Salmonella enterica* subsp. *enterica* serovar Typhimurium infection on Dutch dairy farms. *Prev. Vet. Med.* 54: 157–168.

Vellinga, A. and F. Van Loock. 2002. The dioxin crisis as experiment to determine poultry-related *Campylobacter enteritis*. *Emerg. Infect. Dis.* 8: 19–22.

Wagenaar, J. A., D. J. Mevius, and A. H. Havelaar. 2006. *Campylobacter* in primary animal production and control strategies to reduce the burden of human campylobacteriosis. *Sci. Tech. Rev.* 25: 581–594.

Wagenaar, J. A., W. Jacobs-Reitsma, M. Hofshagen, and D. Newell. 2008. Poultry colonization with *Campylobacter* and its control at the primary production level. In *Campylobacter*, ed. I. Nachamkin, C. M. Szymanski and M. J. Blaser, 667–678. Washington, D.C.: ASM Press.

Wales, A. D., V. M. Allen, and R. H. Davies. 2010. Chemical treatment of animal feed and water for the control of *Salmonella*. *Foodborne Pathog. Dis.* 7: 3–15.

Wales, A., M. Breslin, and R. Davies. 2006. Assessment of cleaning and disinfection in *Salmonella*-contaminated poultry layer houses using qualitative and semi-quantitative culture techniques. *Vet. Microbiol.* 116: 283–293.

Wales, A. D., I. M. McLaren, S. Bedford, J. J. Carrique-Mas, A. J. C. Cook, and R. H. Davies. 2009. Longitudinal survey of the occurrence of *Salmonella* in pigs and the environment of nucleus breeder and multiplier pig herds in England. *Vet. Rec.* 165: 648–657.

Wedderkopp, A., K. O. Gradel, J. C. Jørgensen, and M. Madsen. 2001. Pre-harvest surveillance of *Campylobacter* and *Salmonella* in Danish broiler flocks: A 2-year study. *Int. J. Food Microbiol.* 68: 53–59.

Wedderkopp, A., E. M. Nielsen, and K. Pedersen. 2003. Distribution of *Campylobacter jejuni* Penner serotypes in broiler flocks 1998–2000 in a small Danish community with special reference to serotype 4-complex. *Epidemiol. Infect.* 131: 915–921.

Wedderkopp, A., E. Rattenborg, and M. Madsen. 2000. National surveillance of *Campylobacter* in broilers at slaughter in Denmark in 1998. *Avian Dis.* 44: 993–999.

Wegener, H. C. 2010. Danish initiatives to improve the safety of meat products. *Meat Sci.* 84: 276–283.

Wegener, H. C., T. Hald, D. Lo Fo Wong et al. 2003. *Salmonella* control programs in Denmark. *Emerg. Infect. Dis.* 9: 774–780.

Wheeler, J. G., D. Sethi, J. M. Cowden et al. 2001. Study of infectious intestinal disease in England: Rates in the community, presenting to general practice, and reported to national surveillance. *Brit. Med. J.* 318: 1046–1050.

White, A. P., D. L. Gibson, W. Kim, W. W. Kay, and M. G. Surette. 2006. Thin aggregative fimbriae and cellulose enhance long-term survival and persistence of *Salmonella. J. Bacteriol.* 188: 3219–3227.

Wierup, M. and P. Häggblom. 2010. An assessment of soybeans and other vegetable proteins as source of *Salmonella* contamination in pig production. *Acta Vet. Scand.* 52: 15.

Wilkins, W., A. Rajić, C. Waldner et al. 2010. Distribution of *Salmonella* serovars in breeding, nursery, and grow-to-finish pigs, and risk factors for shedding in ten farrow-to-finish swine farms in Alberta and Saskatchewan. *Can. J. Vet. Res.* 74: 81–90.

Wingstrand, A., J. Neimann, J. Engberg et al. 2006. Fresh chicken as main risk factor for campylobacteriosis, Denmark. *Emerg. Infect. Dis.* 12: 280–284.

Wong, T. L., S. MacDiarmid, and R. Cook. 2009. *Salmonella, Escherichia coli* O157:H7 and *E. coli* biotype 1 in a pilot survey of imported and New Zealand pig meats. *Food Microbiol.* 26: 177–182.

Zaidi, M. B., P. F. McDermott, P. Fedorka-Cray et al. 2006. Nontyphoidal *Salmonella* from human clinical cases, asymptomatic children, and raw retail meats in Yucatan, Mexico. *Clin. Infect. Dis.* 42: 21–28.

Zheng, D. M., M. Bonde, and J. T. Sørensen. 2007. Associations between the proportion of *Salmonella* seropositive slaughter pigs and the presence of herd level risk factors for introduction and transmission of *Salmonella* in 34 Danish organic, outdoor (non-organic) and indoor finishing-pig farms. *Livestock Sci.* 106: 189–199.

Zimmer, M., H. Barnhart, U. Idris, and M. D. Lee. 2003. Detection of *Campylobacter jejuni* strains in the water lines of a commercial broiler house and their relationship to the strains that colonized the chickens. *Avian Dis.* 47: 101–107.

6

Managing Food Safety Risks in the Wild Game and Fish Capture Industries

6.1 Managing Food Safety Risks in Wild Game

6.1.1 Introduction

Meat from wild animals is presumably likely the most original source of animal protein for human consumption (Ahl et al., 2002). Meat-producing large game species are largely ruminant ungulates (hoofed mammals), which are members of the Cervidae (e.g., red deer, roe deer, white-tailed deer) or Bovoidea (e.g., mountain goat, bison, springbok) families. Wild pig or boar (*Sus scrofa*) and rabbits (*Sylvilagus* spp.), hares (*Lepus* spp.), kangaroos (*Macropus* spp.), and crocodilians (*Crocodylus* spp.) are also important sources of game meat (Gill, 2007; McCormick, 2003). Game meats are usually darker, stronger tasting, and often tougher, depending on the type and age of animal (Soriano et al., 2006). Annual consumption of meat from wild and farmed game meat is estimated to be 0.6–1.0 kg per capita in Austria, France, Germany, and Switzerland (Atanassova et al., 2008; Membré et al., 2011). Meanwhile, game meat such as moose, red deer, and reindeer contributed 3.3 kg per capita in Norway (Lillehaug et al., 2005), and ungulates such as wild boars, roe deer, and red deer contributed 1–4 kg per capita in Italy (Ramanzin et al., 2010). Kangaroo meats were harvested and processed in Australia and provided 17,000 tons of meat for consumption. Thirteen thousand tons of meat were exported to the European Union (EU), North America, and Asia (Holds et al., 2008). The risks caused by consumption of game meat are primarily associated with lack of hygiene during processing and with unrecognized zoonotic diseases (Bandick and Hensel, 2011). Feces from wild game could also contaminate surface water, which may then be used as drinking water for humans and/or domestic animals (Lillehaug et al., 2005).

6.1.1.1 Meat from Deer

Venison is a popular game meat in the United States, Central Europe (McCormick, 2003), and United Kingdom (Richards et al., 2011). Consumers are attracted to its tender, low fat content (but favorable fat composition) and high levels of minerals. Deer can be wild, kept in parks, or farmed. Although various species of deer are farmed in different parts of the world, the highest number species of farmed deer are red deer (*Cervus elaphus scoticus*), elk (*Cervus elaphus nelsoni*), and fallow deer (*Dama dama*). In Nordic countries, herding of reindeer are carried out (Hoffman and Wiklund, 2006). Wild deer are usually shot through the chest to enable rapid death, thus minimizing stress and avoiding carcass contamination (Ramanzin et al., 2010). Some hunters prefer to aim for the head or neck to prevent damage to the carcass (Hoffman and Wiklund, 2006; Urquhart and McKendrick, 2006). If the gut is damaged during shooting, this will cause rapid microbial contamination of the carcass (Atanassova et al., 2008; Gill, 2007).

Carcasses of wild deer are generally eviscerated soon after slaughter. Hunters frequently wipe out the inside of the carcass with grass, resulting in grasses becoming attached to the carcass. The head is removed as a trophy and feet are amputated (Coburn et al., 2003, 2005). The likelihood of the gut being damaged during evisceration also varies with the skill of the person performing the task (Gill, 2007). After evisceration, the carcass is transferred to the game processing facility. The time and conditions of transfer are very important to achieve adequate cooling and avoid spoilage especially in warm seasons (Gill, 2007; Paulsen and Winkelmayer, 2004). Delays may occur between the completion of evisceration in the field and placing of the carcass in a refrigerated facility. During the time at ambient temperatures, the flora deposited on the body cavity and cut tissue surfaces will tend to increase (Paulsen and Winkelmayer, 2004). Hunters who handle wild game in the field at times are unaware of the risk of contaminating the meat with foodborne pathogens while dressing, handling, and transporting it. Improper temperature control and inadequate cooking may also contribute to potential foodborne diseases (Figure 6.1).

Salmonella appears to be uncommon in deer. No *Salmonella* were detected in the studies by French et al. (2010), Paulsen and Winkelmayer (2004), Wahlström et al. (2003) (Table 7.1) but an outbreak of *Salmonella* Typhimurium DT104 were found in captive elk (*Cervus elaphus nelsoni*) calves (Foreyt et al., 2001). Most studies on *Salmonella enterica* have demonstrated the rarity of the pathogen in wild game such as red deer and wild boars (Atanassova et al., 2008; Gill, 2007; Paulsen and Winkelmayer, 2004; Wahlström et al., 2003). Table 6.1 shows a summary of some of the microbiological prevalence on various game meats. There seemed to be a low prevalence of foodborne pathogens in deer meat but *T. gondii* were isolated more frequently compared to other pathogens. Wild cervids or deer were found to be carriers of *T. gondii* (Gaffuri et al., 2006; Gauss et al., 2006; Matsumoto et al., 2011; Vikøren et al., 2004). Toxoplasmosis infection in humans was documented after consumption of undercooked raw venison (Sacks et al., 1983) and deer meat and jerky were identified as sources of enterohemorrhagic *Escherichia coli* O157 infections in an outbreak (Keene et al., 1997) and a sporadic case (Rabatsky-Ehr

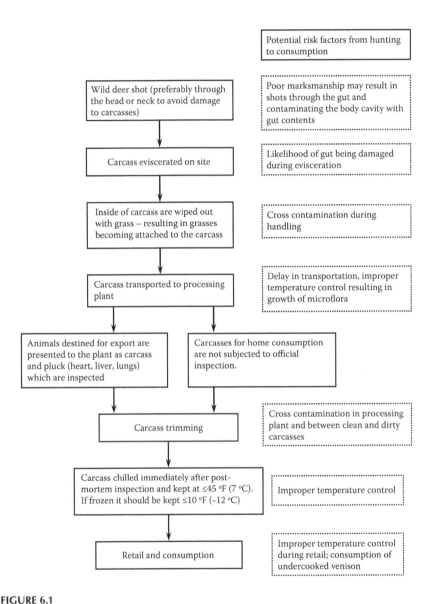

FIGURE 6.1
Potential contamination points during hunting–consumption of wild game meat. (From Coburn, H., E. et al. 2003. Hazards and Risks from Wild Game: A Qualitative Risk Assessment. Centre for Epidemiology and Risk Analysis, Veterinary Laboratories Agency, Weybridge. http://www.foodbase.org.uk//admintools/reportdocuments/660-1-1120_MO1025_Final_Report.pdf (accessed March 2, 2012).; Gill, C. O. 2007. *Meat Sci.* 77: 149–160.; Rabatsky-Ehr, T., et al. 2002. *Emerg. Infect. Dis.* 8: 525–527.).

TABLE 6.1

Prevalence of Microbiological Load on Game Meat

Game	Country or U.S. State	Game	Number of Microb-Positive Samples/Total Number of Samples (%)	Microb	Reference
Kangaroos	Western Australia	Kangaroos (*Macropus fuliginosus*)	23/645 (3.6)	*Salmonella* spp.	Potter et al., 2011
	Western Australia	Kangaroos (*Macropus fuliginosus*)	34/219 (15.5)	*Toxoplasma gondii*	Parameswaran et al., 2009
	Southern Australia	Kangaroos	46/100 (46)	*E. coli*	Holds et al., 2008
		Kangaroos	14/120 (12)	*Salmonella* spp.	Holds et al., 2008
	Queensland, Australia	Kangaroos	9/81 (11)	*Salmonella* spp.	Bensink et al., 1991
Deer	Japan	Sika deer (*Cervus nippon*)	2/107 (1.9)	*T. gondii*	Matsumoto et al., 2011
	Ohio (U.S.)	Captive white-tailed deer (*Odocoileus virginianus*)	1/30 (3.3)	*Escherichia coli* O157	French et al., 2010
			1/30 (3.3)	*Listeria monocytogenes*	
			0/30 (0)	*Salmonella enterica*	
			9/30 (30)	*Yersinia enterocolitica*	
			11/30 (36.7)	*Clostridium difficile*	
	Germany	Roe deer (*Capreolus capreolus*)	4/95 (4)	*Listeria* spp.	Atanassova et al., 2008
		Red deer (*Cervus elaphus*)	3/67 (4)	*Listeria* spp.	
	Italy	Roe deer (*Capreolus capreolus*)	27/207 (13)	*T. gondii*	Gaffuri et al., 2006
		Chamois (*Rupicapra rupicapra*)	5/108 (5)	*T. gondii*	

	Location	n/N (%)	Species	Pathogen	Reference
	Spain	69/441 (15.6)	Red deer (*Cervus elaphus*)	*T. gondii*	Gauss et al., 2006
	Louisiana (U.S.)	1/338 (0.3)	White-tailed deer (*Odocoileus virginianus*)	*Escherichia coli* O157:H7	Dunn et al., 2004
	Austria	0/47 (0)	25 red deer, 18 roe deer, 3 chamois, 1 mouflon	*Salmonella* spp.; *Listeria* spp.	Paulsen and Winkelmayer, 2004
	Sweden	0/285 (0)	Roe deer, fallow and red deer	*Escherichia coli* O157	Wahlström et al., 2003
	Nebraska (U.S.)	4/1608 (0.25)	White-tailed deer (*Odocoileus virginianus*)	*Escherichia coli* O157:H7	Renter et al., 2001
Wild boar	Japan	11/175 (6.3)	Wild boar (*Sus scrofa*)	*T. gondii*	Matsumoto et al., 2011
	Switzerland	18/153 (12)	Wild boar–tonsil samples (*Sus scrofa*)	*Salmonella* spp.	Wacheck et al., 2010
		54/153 (35)		*Yersinia enterocolitica*	
		31/153 (20)		*Yersinia pseudotuberculosis*	
		14/153 (9)		Shiga-toxin producing *Escherichia coli*	
		26/153 (17)		*Listeria monocytogenes*	
	Germany	14/127 (11)	Wild boars	*Listeria* spp.	Atanassova et al., 2008
	Czech Republic	148/565 (26.2)	Wild boar (*Sus scrofa*)	*T. gondii*	Bártová et al., 2006
	Spain	185/507 (38.4)	Wild pig (*Sus scrofa*)	*T. gondii*	Gauss et al., 2005
	Austria	0/3 (0)	Wild boar	*Salmonella* spp.	Paulsen and Winkelmayer, 2004
				Listeria spp.	
	Sweden	0/68 (0)	Wild boars	*Salmonella* spp.	Wahlström et al., 2003

Continued

TABLE 6.1 (*Continued*)
Prevalence of Microbiological Load on Game Meat

Game	Country or U.S. State	Number of Microb-Positive Samples/ Total Number of Samples (%)	Game	Microb	Reference
	Japan	0/46 (0)	Wild boar (*Sus scrofa*)	*Salmonella* spp.; *E. coli* O157	Naya et al., 2003
	Japan	49/131 (37)	Wild boar (*Sus scrofa leucomysta*)	*Yersinia* spp.	Hayashidani et al., 2002
	Japan	2/131 (1)	Wild boar (*Sus scrofa leucomysta*)	*Listeria monocytogenes*	Hayashidani et al., 2002
Crocodiles	Zimbabwe	6/20 (30)	Captive Nile crocodiles (*Crocodylus niloticus*)—fresh samples	*Samonella* spp.	Madsen, 1996
		28/140 (20)	Captive Nile crocodiles (*Crocodylus niloticus*)—frozen samples	*Salmonella* spp.	Madsen, 1996
Birds	Austria	0/74 (0)	Birds (partridges, woodpigeons, quails, 32 pheasants)	*L. monocytogenes*, *Salmonella* spp. and *Campylobacter* spp.	El-Ghareeb et al., 2009
	Slovakia	0/97 (0)	Pheasants	*Salmonella* spp.	Paulsen et al., 2008
	United States	116/128 (91)	Ostrich	*E. coli*	Ley et al., 2001
		0/128 (0)		*E. coli* O157	
		1/152 (0.7)		*Salmonella* spp.	
		19/191 (10)		*Campylobacter* spp.	

et al., 2002). *Escherichia coli* O157 strains had been isolated from deer species (Dunn et al., 2004; Renter et al., 2001). Bullets used to kill wild deer may also result in elevated blood lead concentrations and health risks associated with Pb ingestion (Hunt et al., 2009; Mateo et al., 2007). Hunt et al. (2009) found metal fragments and widespread fragment dispersion in 30 eviscerated deer. After processing the meat, the fragment-containing venison was fed to four pigs to test for bioavailability and four controls received venison without fragments from the same deer. Mean blood lead concentration in pigs peaked at 2.29 µg/dl two days following ingestion of the fragment-containing venison. The level was significantly higher than the controls (0.63 µg/dl). In another study by Iqbal (2008), he found that participants who consumed wild game had 0.30 µg/dl higher blood lead level compared to those who did not consume wild game. The CDC advisory level for intervention in individual children is 10 µg/dl in blood (CDC, 2005) and recently, as little as 2 µg/dl was associated with increased risk of cardiovascular mortality in adults (Menke et al., 2006) and impaired cognitive function in children (Jusko et al., 2008).

Another food safety issue to take into consideration is that deer may play a role as transmitters of *Escherichia coli* O157 strains to cattle by fecal contamination of farmland (Sargeant et al., 1999). Even though cattle are recognized as the primary reservoir for *E. coli*, transmission between cattle to deer, deer to deer, and deer to cattle can still occur. A larger outbreak involved consumption of unpasteurized apple juice, and deer were implicated as the potential source of *E. coli* O157 (Cody et al., 1999).

6.1.1.2 Meat from Wild Boars

Wild boar (*Sus scrofa*) is the ancestor of the domestic pig. It can freely interbreed with domestic pigs. Domesticated pigs that had escaped into the wild led to the establishment of feral pigs (Wilson, 2005). Wild boars hunted in the wild may involve hounds to drive the animals or bring them to bay (Maillard and Fournier, 1995). Carcasses of wild boar are usually skinned in the field or at an abattoir (Gill, 2007). For wild boar carcasses, *Salmonella* were not detected in the meat (Paulsen and Winkelmayer, 2004; Naya et al., 2003; Wahlström et al., 2003) but were recovered from tonsil samples of wild boar (Wacheck et al., 2010). *Yersinia* spp. and *Listeria monocytogenes* (Hayashidani et al., 2002; Wacheck et al., 2010), and *Listeria* spp. (Atanassova et al. 2008) were found in wild boar meat. The meat of wild boars may harbor tissue cysts of *Trichinella gondii* and may represent a vehicle of human toxoplasmosis infection (Bártová et al., 2006; Gauss et al., 2005; Matsumoto et al., 2011) (Table 6.1). With regards to *Trichinella*, this nematode parasite was considered to be transmitted only through consumption of pork (Pozio, 2001), but the role of wild boar meat and other wild animals as sources of infection for humans have been documented. Undercooked wild boar meat had been indicated in a trichinellosis outbreak in Thailand (Jongwutiwes et al., 1998; Khumjui et al., 2008), in Chile (García et al., 2005), in Ontario, Canada (Greenbloom et al., 1997) and Europe (Riccardo et al., 2002). *Trichinella* were also found in bear meat and resulted in a number outbreaks (Gaulin et al., 2000; Schellenberg et al., 2003; McIntyre et al., 2007) and cougar jerky (Dworkin et al., 1996).

6.1.1.3 Meat from Kangaroos

Kangaroo species such as the red (*Macropus rufus*), the eastern grey (*M. giganteus*), the western grey (*M. fuliginosus*), and the euro (*M. robustus*) are usually harvested for meat, skin, and pet food. "Harvesting" is the term used by government agencies and the industry, and "involves the removal of animals that are living in a wild population ... for direct use" (Hercock, 2004). The animals are shot by accredited field processors who operate at night in the Australian outback (Holds et al., 2008). Carcasses are bled and eviscerated in the field and transported directly to the processing premises or are hung in registered wild game cool rooms to await collection. Pathogenic infection in kangaroos is of public health significance due to the kangaroo meat trade in Australia. Bensink et al. (1991) isolated 3 *Salmonella* serotypes (*S.* Bahrenfeld, *S.* Binza, and *S.* Onderstepoort) where nine were positives out of 81 carcasses. *Toxoplasma gondii* was found 15.5% of Australian wild kangaroos (Parameswaran et al., 2009). Potter et al. (2011) later conducted the first extensive study on the prevalence of *Salmonella* infection in wild grey kangaroos in Australia. The overall prevalence of fecal *Salmonella* was 3.6% (Table 6.1) and may have been infected in their natural habitat. Eglezos et al. (2007) isolated *E. coli* from 13.9% and *Salmonella* from 0.84% of samples. Based on the recent studies (Eglezos et al., 2007; Holds et al., 2008; Potter et al., 2011), there are significant microbiological improvements in chilled kangaroo carcasses (as compared to Bensink et al., 1991). The improvements are probably due to the strict regulatory changes (Australian Standard AS 4464-2007; Holds et al., 2008).

6.1.1.4 Meat from Birds

Ostrich meat (*Struthio camelus*) is perceived and marketed as a healthy alternative to other red meats due to its favorable nutritional properties, low cholesterol and intramuscular fat content, and high content of polyunsaturated fatty acids (Fisher et al., 2000; Sales, 1998). South Africa is the main exporter of ostrich meat, exporting approximately 90% of all meat produced (Hoffman and Wiklund, 2006; South African Ostrich Business Chamber, 2012). Ostriches are slaughtered, and the carcasses defeathered, skinned, and eviscerated (Hoffman et al., 2006). The mean numbers of Enterobacteriaceae on dressed carcasses were found ≤10 cm^{-2} (Gill et al. 2000; Severini et al. 2003), while another study detected generic *E. coli* in 91% (116/128) of dressed ostrich carcasses. No *E. coli* O157:H7 were detected. One carcass sample (1/152) was positive for *Salmonella* and *Campylobacter* were detected in 10%

(19/191) of the carcasses (Ley et al., 2001). A study from Austria did not detect *Salmonella* sp., *Campylobacter* sp. or *L. monocytogenes* from hunted wild birds (El-Ghareeb et al., 2009; Table 6.1). However, birds from pheasantries may have a higher prevalence of *Salmonella* than pheasants in the wild (Coburn et al., 2003).

6.1.1.5 Meat from Crocodiles

The main crocodile industries exist in Africa, particularly Zimbabwe and South Africa where the Nile crocodile (*Crocodylus niloticus*) is reared (Madsen et al., 1992), while Australia and Asia mainly farm saltwater crocodiles (*C. porosus* and *C. johnstoni*). Nile crocodiles are also reared in Kenya, Tanzania, Israel, Indonesia, France, Japan, and Spain and the United Kingdom (license for farming in the United Kingdom was awarded for the first time in 2006) (EFSA, 2007). The crocodile industry is based on a mixture of farming or ranching. Only farmed or ranched crocodiles are considered and not wild ones ("bush meat"). Farming refers to captive breeding, while ranching is dependent on the wild population. In ranching, eggs or hatchlings are collected from natural habitats and reared to harvest size in captivity (Magnino et al., 2009). Crocodiles are primarily reared for their skins while meat is usually a by-product. *Salmonella* was isolated in 30% of fresh meat samples and 20% of frozen meat samples that were to be processed for human consumption from captive Nile crocodiles (Table 6.1). It was suggested that the presence of *Salmonella* in meat samples may be due to skin surface contamination originating from fecal-polluted rearing water ponds combined with excessive handling procedures during skinning. In a separate study, *Salmonella* was detected in pond water of crocodile farms in Zimbabwe (Madsen, 1994). Contamination of the meat is likely because the skin is valuable and must be removed with care. Since the skin does not peel off easily, crocodiles must be skinned on a flat surface, thus providing greater opportunity for cross-contamination of the meat (Madsen et al., 1992).

From the above review, it can be seen that the presence of pathogenic bacteria in wild game varies across different locations. This could be attributed to the particular geographical area where animals are free living species and do not share pastures with domestic animals. Similarly, transmission can occur between wild and domestic animals when both groups are sharing the same habitat.

6.1.2 Intervention Strategies

Game meat is becoming more popular and consumers are more critical of the safety and quality of wild game. This reiterates the importance of the hygienic and microbiological status of wild game meat. In the European Union, Regulations (EC) No. 852/2004, 854/2004, and 178/2002 also apply to wild game meat. Hunters who sell game to wholesalers or game processing plants are considered as food business operators and are responsible for food safety. They also have to ensure traceability of the meat (i.e., prove the source of game and to whom they sold it). According to Regulation (EC) No. 853/2004 for wild game meat, at least one person in the hunting team who hunts wild game in order to place it on the market for human consumption must have sufficient knowledge of its pathology and handling after hunting, including the ability to undertake an initial examination of wild game on the spot.

6.1.2.1 Skills of Game Hunters

Hunting methods may differ between countries. For example, some may make use of dogs to drive the animals toward the hunters, and some hunters may stalk the quarry or wait for it from vantage points. Hunting wild game with dogs is much more stressful for the animals (Bateson and Bradshaw, 1997; Bradshaw and Bateson, 2000). High concentrations were found of cortisol, associated with extreme physiological and psychological stress in the hunted deer. Moreover, the chase took place over a distance of 19 km (Bateson and Bradshaw, 1997). The stress associated with hunting may compromise meat quality.

Game hygiene as well as the quality of shot game animals is dependent on the attitude of the hunter, health of the animal before being shot, target positions, and behavior of the animal after shooting (Atanassova et al., 2008). The modes of killing (in Central Europe) for large game are free bullet in head, neck, or anterior chest, while for small game the mode is with multiple shot pellets. Professional hunters prefer to shoot deer in the head, neck, or in the heart to minimize damage to carcasses (Urquhart and McKendrick, 2006). A shot in the heart is quickly fatal causing minimal destruction of the meat; little microbial stress and minimal fleeing distance thus resulting in better microbiological quality (Atanassova et al., 2008). The researchers also investigated the connection between skillful and less skillful hunters and the occurrence of Enterobacteriaceae. In the case of expertly shot wild game, Enterobacteriaceae were found in 11% to 25.6% of the carcasses. Meanwhile, in nonexpert shots, Enterobacteriaceae were found in 31.3% to 51.7% of the shot game. This indicates the importance of killing methods for hygiene quality of game meat. Poor placement of shots, efflux of ingesta or feces during evisceration, and delayed or insufficient cooling will compromise the microbiological condition of meat (Paulsen, 2011).

Highly elevated Pb levels in game meat can be easily reduced and addressed with suitable legislation banning the use of Pb-based ammunition in game hunting. Lead-free copper bullets are now widely available and can be utilized to reduce Pb contamination (Taggart et al., 2011). Copper is less toxic than lead with little or no fragmentation (Hunt et al., 2006). Many countries such as Russia, North America, United Kingdom, Japan, Israel, Ghana, South America, Malaysia, and some countries in the European Union have implemented total or partial bans on the use of Pb shot for hunting (Mateo et al., 2007).

6.1.2.2 Training of Game Hunters

Hunters should be made aware of the microbiological risks—especially *T. gondii* infections in wild boars. Precautions should be taken when hunting game animals, and any game meat should be cooked thoroughly before consumption. Training should be provided to hunters (Regulation EC No. 853/2004) in the following subjects:

- The normal anatomy, physiology and behavior of wild game;
- Abnormal behavior and pathological changes in wild game due to diseases, environmental contamination, or other factors which may affect human health after consumption;
- The hygiene rules and proper techniques for the handling, transportation, evisceration, etc., of wild game animals after killing; and
- Legislation and administrative requirements.

The authorities should encourage hunter associations to promote this kind of training. In some countries, where there is no "trained" hunter, a postmortem inspection must be carried out at the slaughterhouse. In addition to a visual examination, a trichinoscopic examination must be carried out on animal species susceptible to *Trichinella* infections (Regulation EC 2075/2005). It is also necessary to examine against foreign bodies (e.g., lead shots and projectiles) in the meat (Casoli et al., 2005). On the other hand, hunters who hunted game meat for their families or acquaintances are usually not subjected to stringent meat inspection (Ramanzin et al., 2010). It is imperative that the public, especially among communities that consumed a higher proportion of game meat), are educated on the appropriate hunting and slaughtering of game while in the fields. For example, in the state of Montana, the Department of Fish, Wildlife and Parks and Montana State University (extension service) published and circulated pamphlets to hunters on the dangers of eating improperly cooked bear meat. Another important factor in the epidemiology of trichinellosis in wild game populations is the habit of hunters who tend to leave offal and carcasses of game in the forests where they serve as a source of infection for other animals. By enforcing safe disposal of offal from hunted wild boars, this may reduce the incidence of trichinellosis in wild game (Murrell and Pozio, 2000).

6.1.2.3 Handling of Wild Game after Killing

Proper handling of game carcasses begins in the field after killing. Immediate evisceration of the intestines/entrails is carried out. If any of the internal organs smell offensive or exhibit discharge or blood in the muscle, the flesh is unfit for consumption. The abdominal cavity should be cleaned, dried, and cooled to <41°F (5°C) until the meat is processed (Rabatsky-Ehr et al., 2002). In Paulsen and Winkelmayer (2004), the authors investigated the influence of an antemortem precooling phase on the microbial quality of carcasses. The temperature differences during different seasons also affect the microbial load on carcass surfaces. Microbial loads tend to be lower during winter compared to the summer months even when evisceration was performed in a "clean" way, and no contamination was visible. Regardless of seasons, an effective cooling regime will ensure optimum meat quality from game species.

Avagnina et al. (2012) recorded high levels of contamination in animals eviscerated and sampled within 30–60 minutes after shooting and in animals eviscerated and sampled more than 180 minutes after shooting. This indicates that besides the influence of the time elapsed between shooting and evisceration, handling and harvesting practices also play an important role in determining carcass microbial loads in game meat. The time between cooling and cutting of the carcass is important since this is a period when microbes grow and meat spoils. When temperatures are high, skinning is recommended to allow for a more rapid removal of body heat (Field, 2004). In the framework of Regulation 852/2004, skinning and butchering are performed at approved game-handling establishments under veterinary control with formal Hazard Analysis and Critical Control Point (HACCP) principles. This can ensure that the hygiene of carcass processing and the cool chain are maintained (Regulation (EC) No. 852/2004).

In South Africa and Zimbabwe where farmed crocodiles are operated in commercial operation, it is necessary to control *Salmonella* spp. contamination to crocodilian meat. It was suggested that a system for skinning and dressing of crocodiles in a hanging position may prove useful where it can preserve the integrity of the much prized leather while maintaining hygienic practices. Furthermore, dipping the meat in a 30 ppm chlorine solution for 10 minutes reduced surface contamination by approximately 90% (Madsen et al., 1992).

6.1.2.4 Traceability

Traceability is ensured from the time the carcass is dressed in the field and approved by a "trained" person who has to attach to the carcass a signed declaration (Ramanzin et al., 2010). Farms that supply game animals for export must register with the Controlling Authority (a body under a law of the country or a province that has statutory responsibility for game meat hygiene) prior to being allowed to export the game meat. If the farm complies with all the requirements, then it is issued with a registration number (renewable annually) that is linked to all the game meat from the farm. This facilitates traceability of the meat back to the farm of origin and a detailed knowledge of the health status of the animals (Hoffman and Wiklund, 2006).

6.1.3 Case Study of Outbreak of Trichinellosis due to Consumption of Bear Meat

In 2005, an outbreak of trichinellosis in Victoria, Canada, affected 26 probable cases due to consumption of black bear meat. The bear was shot and the meat was frozen for at least 3 days before being barbecued, fried, or stewed at three separate events. However, freezing is not adequate to prevent *Trichinella* transmission. Meat should be cooked thoroughly to achieve an internal temperature of 171°F (77°C). It was possible that other foods and surfaces were contaminated during food preparation since individuals who consumed the very well cooked (stewed) bear meat were also infected (McIntyre et al., 2007; Schellenberg et al., 2003). Outbreaks of trichinellosis in Canada were due to *T. nativa*, which is generally found in hosts from the arctic and subarctic regions (e.g., black bears, grizzly bears, polar bears, and walruses) and is resistant to freezing (Appleyard and Gajadhar, 2000). Meanwhile in the year 2000, another similar outbreak occurred in two Northern Saskatchewan communities where a total of 78 individuals had consumed bear meat infected with *T. nativa*. The 31 confirmed cases were more likely to have eaten dried bear meat while noncases indicated that they ate only boiled bear meat. The former meat was dried/smoked over an open fire for 2–3 days and probably did not generate enough internal heat to kill the cysts of *T. nativa*. Hunters who shoot wild game in arctic and sub-arctic areas of North America must be made aware of the risks of development of trichinellosis in inadequately cooked meat (Schellenberg et al., 2003).

6.2 Managing Food Safety Risks in the Fishery Industries

6.2.1 Introduction

The microbial status of marine seafood is closely related to the environmental conditions and water quality such as water temperature, salt content, distance between catch area, and polluted areas (human and animal feces). Other factors include natural occurrences of bacteria in water, mode of feeding, harvesting season, methods of catch and chilling conditions. There will be differences in microbial status according to harvesting area, that is, seafood harvested from deep sea, near to coastal areas, or rivers and lakes. Generally, risks of seafood harvested from unpolluted marine waters are low (Feldhusen, 2000). Seafood is responsible for an important proportion of foodborne illness and outbreaks in the United States, European Union, and worldwide. Seafood includes mollusks (e.g., oysters, clams, and mussels), finfish (e.g., tuna, salmon, barracuda, grouper), marine mammals (whale and seal), fish eggs (roe), and crustaceans (e.g., shrimp, crab, and lobsters; Iwamoto et al., 2010). In this section, we will cover marine finfish harvested from the wild and mollusks. Farmed finfish (e.g., salmon) and crustaceans (e.g., shrimp, which are mostly farmed today) will be discussed in the next chapter.

Foodborne pathogens have been estimated to cause as many as 9.4 million illnesses, 55,961 hospitalizations, and 1,351 deaths in the United States each year (Scallan et al., 2011). Butt et al. (2004a) estimated 10–19% of these illnesses were attributed to seafood consumption. Meanwhile, Adak et al. (2005) estimated that between 1996 and 2000, there were more than one million cases of indigenous foodborne disease per year, resulting in over 20,000 hospitalizations and more than 600 deaths in England and Wales. Shellfish was the riskiest item to eat in terms of number of cases per serving, and was estimated at over six times the risk of poultry and nearly 27 times the risk of red meat. Finfish are less likely to be associated with infectious illnesses because they are often eaten well cooked (with the exception of sashimi or raw fish). Mollusks are more frequently marketed and eaten raw or partly cooked, hence increasing the risk of infectious illnesses caused by pathogens that would have been killed or inactivated by heat (Butt et al., 2004a).

Most of the reported seafood-associated outbreaks were attributed to eating contaminated shellfish, of which *Vibrio parahaemolyticus* and viruses were the most common agent in these outbreaks. Shellfish contributed the highest number of breakdowns in the seafood supply chain in the European Union, which were mostly of viral origin (Doyle et al., 2004; Guillois-Becel et al., 2009; Malfait et al., 1996). Oyster harvesting areas are prone to human sewage discharge. Since oysters are filter-feeders, as water flows through them, they ingest and concentrate all particulate matter in the water, including pathogenic bacteria and viruses (Martinez-Urtaza et al., 2003). *Vibrio parahaemolyticus* has been recognized as a major cause of gastroenteritis associated with the consumption of seafood, especially contaminated oysters (Potasman et al., 2002). Shellfish are more likely to transmit viruses because as filter-feeders, they concentrate viral agents from the surrounding waters and are often eaten raw or undercooked (Wallace et al., 1999). Meanwhile, from 1992 to 2001, 31.1% of the 5770 foodborne outbreaks in China were caused by seafood (Liu et al., 2004). Hepatitis A virus also caused the largest-ever shellfish-associated epidemic in China in 1988 where 292,301 cases and 32 deaths were reported due to consumption of raw clams. Hepatitis A virus was traced back to clams from the markets and catching area (Halliday et al., 1991). In 1993–1996, five large oyster-related outbreaks due to Norwalk-like viruses occurred due to oyster contamination prior to harvest. All five outbreaks identified human sewage as sources of contamination. The contaminations were mostly due to practices of dumping sewage overboard and a malfunctioned sewage facility of an oil rig (Berg et al., 2000). The presence of biotoxins is another major health concern in many parts of the world. Ciguatera is a disease associated with tropical fish. The toxin is produced by dinoflagellates, and the fish become toxic through food chain magnification. Unfortunately, there are no signs to warn consumers when a fish is contaminated with ciguatoxin.

6.2.2 Bacterial Pathogens in Seafood

6.2.2.1 *Vibrio spp.*

Vibrios are a frequent isolate from seafood, particularly shellfish. *Vibrio cholerae*, *V. parahaemolyticus*, and *V. vulnificus* are universally recognized as important human pathogens. *V. parahaemolyticus* is the most commonly isolated "noncholera" vibrio (Feldhusen, 2000). Consumption of raw or undercooked seafood, particularly shellfish contaminated with *Vibrio parahaemolyticus* may lead to the development of acute gastroenteritis characterized by diarrhea, headache, vomiting, nausea, abdominal cramps, and low fever (Su and Liu, 2007). *Vibrio parahaemolyticus* was the leading cause of food poisoning in Japan with 4225 outbreaks and 118,268 cases from 1981 to 1995 (Lee et al., 2001) and 1710 outbreaks and 24,373 cases between 1996 and 1998 (IDSC, 1999). In the United States, raw oysters are the most important vehicles for transmission of *V. parahaemolyticus* (Drake et al., 2000; McLaughlin et al., 2005; Su and Liu, 2007). Outbreaks of *V. parahaemolyticus* in raw oysters were also reported in Chile (Cabello et al., 2007) and Brazil (Leal et al., 2008). In these outbreaks, rising seawater temperatures and intertidal exposure to solar radiation during ebb and low tides (reaching temperature of 86°F [30°C]) can increase the concentration of *V. parahaemolyticus* in shellfish and the ocean, hence increasing the risk for human infection after consumption (Cabello et al., 2007; McLaughlin et al., 2005). Disease caused by *V. parahaemolyticus* is rarely reported in Europe but had occurred in Spain (Lozano-León et al., 2003; Martinez-Urtaza et al., 2005) and Italy (Ottaviani et al., 2008). *Vibrio* spp. are particularly prevalent in tropical waters and can be isolated in temperate zones during the summer months. Kaneko and Colwell (1973) could not detect vibrios in the waters of Chesapeake Bay during the winter months. However, when water temperatures rose above 57°F (14°C), vibrios were released from the sediments, attached to zooplankton, and proliferated as the temperature increased. In a more recent study, the densities of *V. parahaemolyticus* were also found to be positively correlated to water temperatures, with higher densities occurring in summer (Duan and Su, 2005).

6.2.2.2 *Salmonella spp.*

Heinitz et al. (2000) demonstrated the presence of *Salmonella* in a variety of fish and shellfish, including seafood intended for consumption without further preparation. Sewage contamination of shellfish harvest beds led to large shellfish-associated outbreaks of *Salmonella* serotype Typhi infections (Iwamoto et al., 2010). In a study by Brands et al. (2005), *Salmonella* was isolated from oysters from each coast of the United States (West, East, and Gulf coasts). Ninety-three of 1296 oysters examined were positive for *Salmonella* spp., where *Salmonella enterica* serovar Newport was the predominant serotype isolated from oysters. Shellfish are known carriers of viral and bacterial pathogens. Heinitz et al. (2000) detected *Salmonella* in 1.2% of the local shellfish in the United States. *Salmonella* prevalence in seawater and seafood is influenced by climatic conditions—especially different seasons (Brands et al., 2005), wind, rainfall (Martinez-Urtaza et al., 2004; Setti et al., 2009), and stormwater (Haley et al., 2009). There was a lower incidence of *Salmonella* isolated from oysters during winter months compared to summer. Differences in *Salmonella* isolation between summer and winter may be due to the temperature of the water, with colder waters reducing the presence of bacteria while warmer water allows increased bacterial survival (Rhodes and Kator, 1988). Animals are also more active during the summer months, and their increased defecation into rivers that feed oyster bays may result in higher numbers of *Salmonella* (Clegg et al., 1983). Rainfall has been identified as the universal environmental driver (Simenthal and Martinez-Urtaza, 2008) with storm waters participating in the transport of *Salmonella* from its source points to marine environments (Haley et al., 2009; Martinez-Urtaza et al., 2004).

Two successive *Salmonella* Montevideo outbreaks occurred among guests attending two different social functions, provided by the same catering company. Improperly prepared chilled, boiled salmon was found as the most likely vehicle of infection in both outbreaks (Cartwright and Evans, 1988). Another salmonella outbreak of *Salmonella paratyphi* B infection in the United Kingdom was associated with an infected handler from a fish and chips shop (Francis et al., 1989). Both outbreaks were indicative of failure in adhering to safe food-handling practices. Adequate cooking kills most pathogens; but unlike other foods such as poultry that are usually fully cooked, fish and seafood is often consumed raw or prepared in other ways that may be insufficient to kill pathogens (Iwamoto et al., 2010).

6.2.2.3 *Listeria monocytogenes*

Listeria monocytogenes is considered a ubiquitous organism and can be found in soil, foliage, and feces of animals and humans. The bacterium is able to grow under both aerobic and anaerobic conditions and at temperatures normally found during refrigeration. This makes it challenging for fish processing, which operates under cool conditions, since *L. monocytogenes* can proliferate under chilled conditions. The general tendency is that the fraction of positive samples and the number of bacteria in these samples increases during processing of product with minimal or no preservation. Studies found that the prevalence of *L. monocytogenes* in raw fish were low, ranging from 0 to 1% (Autio et al., 1999) to 10% (Johansson et al., 1999). The reported incidence of *Listeria* spp. in seafood from tropical waters is also low (Karunasagar and Karunasagar, 2000).

Thus, the raw materials generally contain less *L. monocytogenes* than the ready-to-eat product, for example, smoked salmon (Lunestad and Rosnes, 2008). Contaminated seafood such as smoked mussels and cold-smoked rainbow trout has been implicated as sources of listeriosis outbreaks (Brett et al.,

1998; Ericsson et al., 1997). Chen et al. (2010a) did not detect *L. monocytogenes* on catfish skins and intestines, but were found in chilled and unchilled fillets. *Listeria* spp. were also detected on fish contact surfaces. These results coincide with those given by Autio et al. (1999) and Dauphin et al. (2001) who concluded that processing plants (for smoked salmon) are particularly susceptible to contamination due to the difficulties in efficient cleaning and disinfecting. Autio et al. (1999) found only one sample among 60 fish from the processing plant contained *L. monocytogenes*. However, the number of fish samples positive for *L. monocytogenes* increased after the brining stage. This study strongly reiterates the fact that *L. monocytogenes* contamination in the processed fillets originated from the processing environment rather than from the fish. Smoked salmon processing plants involves a lot of handling by workers (filleting, salting, drying, smoking, trimming, and packaging) and the use of complex technical equipment (e.g., mechanical brine injector) and may facilitate colonization of *L. monocytogenes* in the processing environment and spread to the product (Rørvik, 2000).

6.2.2.4 Aeromonas spp.

Aeromonas spp. is an environmental microorganism found in aquatic environments and can be sporadically transmitted to humans (Castro-Escarpulli et al., 2003). *Aeromonas* (particularly *A. hydrophila*) has the status of a foodborne pathogen and attracted attention due to its ability to grow at low temperatures (as low as 28°F [–2°C]; Daskalov, 2006). *Aeromonas* spp. had been implicated in foodborne outbreaks due to contaminated shrimps (Altwegg et al., 1991). *Aeromonas* spp. were isolated from frozen fish intended for human consumption (Castro-Escarpulli et al., 2003), hot- and cold-smoked fish, and gravad salmon (Gobat and Jemmi, 1993), fish from retail markets (Radu et al., 2003), and fish and prawns from a seafood market (Vivekanandhan et al., 2005). In addition, *Aeromonas* spp., particularly *A. salmonicida*, is an important fish pathogen. *Aeromonas salmonicida* is a causative agent for furunculosis or the appearance of boil-like sores or lesions on fish.

6.2.2.5 Viruses in Seafood

Norovirus and Hepatitis A virus are two of the most important foodborne viruses. Such viruses can contaminate seafood through fecal contaminated seawater or during seafood processing through inadequate hygiene practices of operators or systems (Lees, 2000). Of the various seafood species, bivalve mollusks is the constant vehicle for transmission of these viruses. Since mollusks obtain food by filtering a large volume of water, during the process they accumulate particulate matter including viruses. This process resulted in higher concentration of viruses within the shellfish. For example, one study showed that fecal contaminants were 3 to 62 times greater in an oyster as compared to the surrounding water (Burkhardt and Calci,

2000). Contamination occurs particularly when shellfish were harvested from coastal waters which are close to urban sewage treatment facilities. Viruses can survive for weeks (Callahan et al., 1995) to months especially in lower temperature 39°F (4°C) and reduced UV light conditions (Gantzer et al., 1998). One of the largest Hepatitis A outbreaks occurred in Shanghai, China, in 1988, where almost 300,000 people were infected after eating clams harvested from an area contaminated with untreated sewage from the surrounding residential area and untreated effluent from the fishing vessels in the catching area (Halliday et al., 1991). Standard guidelines are important in the regulation of shellfish harvesting and routine monitoring of waters, but it is when authorized shellfish harvesting areas decrease that unethical activities such as illegal harvesting from polluted or restricted area become problematic (Jones et al., 1991).

6.2.2.6 *Parasites in Seafood*

Consumption of raw and undercooked seafood is associated with parasitic infections, particularly with anisakids and cestodes. Anisakids are a group of nematodes that include species of the genera *Anisakis, Pseudoterranova, Contracaecum,* and *Hysterothylacium,* which infect fish (Herrero et al., 2011). *Anisakis* spp. (herring or whale worm) commonly occurs in most commercially important fish species. The best known species is *A. simplex,* followed by *Pseudoterranova decipiens* (Herreras et al., 2000). In fish, *Anisakis* larvae are encapsulated as flat tight spirals embedded in or on the visceral organs and peritoneum. Some larvae may migrate from the viscera into the flesh (Levsen and Lunestad, 2008). After ingestion, a viable juvenile anisakid may bore into the wall of the gastrointestinal tract causing an immune reaction. Outbreaks of anisakiasis occur especially in countries with high consumption of raw or undercooked fish (Puente et al., 2008).

Diphyllobothrium latum (broad tapeworm), which causes diphyllobothriosis usually occurs in areas where fish are eaten raw or undercooked. The worm is a serious human pathogen in Eastern Europe and Russia. It has also occurred regularly in North America, especially around the Great Lakes and Alaska (Levsen and Lunestad, 2008). *Diphyllobothrium latum* competes for Vitamin B_{12} in the human host, leading to pernicious anemia and may grow up to 10 m in length (Butt et al., 2004b). From July 2006 to September 2010 in the state of Missouri, there were nine reported cases of paragonimiasis. Paragonimiasis is a type of parasitic disease caused by *Paragonimus* trematodes, commonly known as lung flukes. Humans can become infected by eating raw or undercooked crayfish or freshwater crabs that harbor the parasites. All affected patients reported having eaten raw or undercooked crayfish from rivers in Missouri while on canoeing or camping trips (CDC, 2010). Once ingested, larvae will penetrate the peritoneal cavity and move across the diaphragm into the pleural cavity. Finally, the parasites migrate into the lung parenchyma where they reach maturity and form solid worm

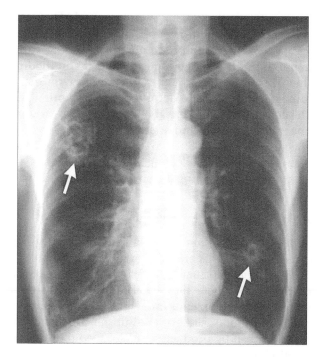

FIGURE 6.2
Chest radiograph showing typical cavitating lesions caused by *Paragonimus westermani* infection. The patient ate sashimi of wild boar meat. (From Nawa, Y., C. Hatz, and J. Blum. 2005. *Clin. Infect. Dis.* 41: 1297–1303.)

cysts, hence the name lung flukes. Typical nodular cavitating lesions caused by *P. westermani* infection is shown (Figure 6.2).

6.2.3 Fish Poisoning

Ciguatera fish poisoning is a type of foodborne disease that results from eating predatory ocean fish contaminated with ciguatoxins. The condition is endemic in tropical and subtropical regions of the Pacific basin, Indian Ocean, and Caribbean. In the United States, 5–70 cases per 10,000 persons are estimated to occur yearly due to ciguatera fish poisoning (Gessner and Mclaughlin, 2008). In 2007, a cluster of nine cases of ciguatera fish poisoning occurred in North Carolina. The patients reported having eaten amberjack bought from a fish market (CDC, 2009). Carnivorous tropical and subtropical fish such as barracuda, amberjack, red snapper, and grouper can become contaminated with ciguatoxins by feeding on plant-eating fish that have ingested *Gambierdiscus* spp., a group of dinoflagellates commonly found in coral reef waters (Lewis, 2001). Spoilage of fish that have been caught is not a factor in toxin development and cooking does not deactivate the toxin (CDC, 2009). Meanwhile, histamine is related to scombroid poisoning and is caused by abusive handling (Huss et al., 2000). Symptoms include nausea, vomiting, diarrhea, oral burning sensation or peppery taste, hives, itching, and red rash (Taylor et al., 1989). Severity of scombroid poisoning varies considerably with the amount of histamine ingested and the individual's sensitivity to histamine. Histamine is formed by certain bacteria that possess the enzyme histidine decarboxylase to decarboxylate the amino acid histidine. Scombroid fish such as tuna, mackerel, bonito, and saury contain high levels of free histidine in their muscle and are often implicated in scombroid poisoning incidents (Chen et al., 2008; Taylor et al., 1989). Several nonscombroid fish such as mahi-mahi, bluefish, herring, and sardine were also implicated in incidents of scombroid poisoning (Chen et al., 2010b; Tsai et al., 2005).

6.2.4 Chemical Contaminants in Seafood

Toxic chemical contaminants such as heavy metals (mercury, Hg; cadmium, Cd; lead, Pb; arsenic, As) and persistent organochlorine pollutants (POPs) such as dioxins and polychlorinated biphenyls (PCBs) are major concerns in aquatic environments where they bioaccumulate in fish and shellfish. Table 6.2 shows examples of environmental contaminants, particularly POPs, found in wild salmon. Undesirable substances such as mercury are commonly present in fish by-products used for feed (Berntssen et al., 2010; Chou, 2007). Methyl-mercury is efficiently accumulated in fish muscle and is the dominant form of mercury in fish, while fish meal is the main source for methyl-mercury in fish feeds (Håstein et al., 2006). Bioaccumulation of methyl-mercury in fish depends on the trophic level, age, or length of fish (Zhang and Wong, 2007). Particularly fishery species like swordfish, longtail tuna, thornback ray (Storelli et al., 2012), and black scabbard fish (Afonso et al., 2007) accumulate high Hg concentrations in muscle tissues while cuttlefish contained the highest Cd concentration (Storelli et al., 2012). Crabs and lobsters hepatopancreas were also found to be particularly high Cd concentration (up to 40 mg/kg) (Barrento et al., 2008, 2009; Noël et al., 2011). The main dietary source of arsenic in humans is from seafood, but this is not of concern from a food safety perspective since the predominant form is organic arsenic, which is considered nontoxic (Maage et al., 2008).

Persistent organic pollutants (POPs) such as polychlorinated dibenzo-*p*-dioxins (PCDDs), polychlorinated dibenzofurans (PCDFs), and polychlorinated biphenyls (PCBs) represent a group of highly toxic substances accumulated in the tissues of marine organisms and conveyed through the food chain to humans. PCDD/Fs and PCBs are released into the environment from sources such as waste incineration and chemical manufacturing. PCBs in sediments may be taken up by bottom-dwelling organisms and bioaccumulate as it moves up the food chain. As fish eats other fish or bottom-dwelling organisms, they take on the body burden of PCBs present in their prey. Fish are able to metabolize some PCBs while for those that are not metabolized or excreted, will accumulate in fatty tissues (Fensterheim, 1993).

TABLE 6.2

Contaminants Found in Wild Salmon

Samples	Contaminants				References
	PCBs	PBDEs	PCDD/Fs	DDT	
Wild salmon (Chile)	934–14.1 ng g⁻¹ wwt	0.4 ng g⁻¹ wwt			Montory et al., 2010
Wild salmon (Poland)		2.5 ng g⁻¹ wwt			Szlinder-Richert et al., 2010
Wild salmon (North Atlantic)	3262 ng g⁻¹ lipid	263 ng g⁻¹ lipid		4063 ng g⁻¹ lipid	Svendsen et al., 2007
Wild salmon (U.S.)	4.0 ng g⁻¹ wwt	0.3 ng g⁻¹ wwt			Hayward et al., 2007
Wild juvenile salmon (U.S.)	1300–14,000 ng g⁻¹ lipid			1800–27,000 ng g⁻¹ lipid	Johnson et al., 2007
Wild salmon	3 ng g⁻¹ wwt		0.03 pg g⁻¹ wwt		Mozaffarian and Rimm, 2006
Chinook salmon (U.S.)[a]	12.6 µg kg⁻¹	1.8 µg kg⁻¹			Stone, 2006
Chinook salmon (U.S.)[a]	49.26 ng g⁻¹ wwt				Missildine et al., 2005
Wild salmon (U.S.)		0.049–0.112 ng g⁻¹ wwt			Hites et al., 2004
Wild coho and Chinook salmon	1450 ng g⁻¹ wwt	80.1 ng g⁻¹ wwt			Manchester-Neesvig et al., 2001
Wild salmon (Japan)		703 pg g⁻¹ wwt			Ohta et al., 2002
Wild salmon (Canada)	5.30 ng g⁻¹ wwt	0.18 ng g⁻¹ wwt			Easton et al., 2002

Note: PCB₇—Polychlorinated biphenyl 28; 52; 101; 118; 138; 153; 180. Due to the large number of individual congeners (209), a complete analysis of PCB is difficult; hence scientific consensus measure seven individual congeners known as PCB₇. PBDE—Polybrominated diphenyl ethers.

[a] Salmon released from hatcheries and caught when salmon returns for spawning.

6.2.5 Intervention Strategies

6.2.5.1 Harvesting Practices

The control of food safety risks associated with bivalves requires Hazard Analysis Critical Control Point (HACCP) procedures and water environmental quality management of growing and harvesting areas and postharvest product processing (WHO, 2010). Contamination may be reduced by heat treatment (cooking) or by extending the natural filter-feeding processes in clean seawater to purge out microbial contaminants. This can be done in tanks (depuration) or in the natural environment (relaying). Both the European Union and the United States use fecal indicators (*E. coli* or fecal or total coliform) to monitor the quality of harvesting area. The fecal monitoring will determine the appropriate treatment in accordance with the level of contamination and the prescribed statutory standards. In the European Union, fecal indicators (*E. coli*) are measured in shellfish flesh and intravalvular liquid, whereas in the United States, indicators are measured in the shellfish growing waters. In the United States, either fecal coliform or total coliform may be used for monitoring. U.S. Food and Drug Administration (USFDA) "approved" and EU "Category A" standards describe the cleanest growing areas from which shellfish can be taken for direct human consumption without further processing.

Shellfish from the EU category A area must contain less than 230 *E. coli* in 100 g of shellfish flesh. The approved areas in the United States must comply with a total coliform geometric mean (GM) of 70 per 100 ml water with not more than 10% of samples exceeding 230 per 100 ml. Alternatively, they can comply with a fecal coliform GM of 14 per 100 ml water with not more than 10% of samples exceeding 43 per 100 ml (Table 6.3). Shellfish harvested from polluted areas (category B and C) are allowed when shellfish undergo treatments before being commercialized. Bivalve shellfish growing areas exceeding these prescribed levels of contamination, or areas for which harvesting area survey and classification has not been conducted, cannot be harvested for human consumption in either the United States or European Union (Lees, 2000; WHO, 2010).

A time-temperature matrix (Table 6.4) was developed by the National Shellfish Sanitation Program (NSSP, 2009) to limit the exposure time of oysters to an elevated temperature limit, which in turn limits the growth of vibrios in shellfish. For example, oysters harvested for raw consumption and which were subjected to an average monthly maximum air temperature of $\geq 81°F$ ($\geq 27°C$) must be cooled down to $50°F$ ($10°C$) within 10 hours. Oyster beds are closed during the warm summer months and at other times when heavy rains cause an influx of potentially contaminated water (Andrews, 2004).

Mussel Watch biomonitoring programs were organized to monitor chemical contamination and environmental quality of coastal areas, especially where production of shellfish takes place (Amiard et al., 2008). The Center for Disease Control and Prevention (CDC) also implemented CaliciNet, the

TABLE 6.3

EU and USA Legislative Standards for Live Shellfish

Shellfish Treatment Required	US FDA Classification	Microbiological Standard per 100 ml Water	EU Classification	Microbiological Standard per 100 g Shellfish
Not required	Approved	GM < 14 FCs and 90% < 43 FCs	Category A	All samples < 230 *E. coli* or All samples < 300 FCs
Depuration or relaying	Restricted	GM < 88 FCs and 90% < 260 FCs	Category B	90% < 4600 *E. coli* or 90% < 6000 FCs
Protracted relaying (> 2 months)	US FDA did not incorporate an equivalent to EU category C		Category C	All samples < 60,000
Harvesting prohibited	—	If above levels are exceeded	—	If above levels are exceeded

Source: Lees, D. 2000. Viruses and bivalve shellfish. *Int. J. Food Microbiol.* 59: 81–116; WHO. 2010. *Safe Management of Shellfish and Harvest Waters*, ed. G. Rees, K. Pond, D. Kay, J. Bartram, and J. Santo Domingo, London: IWA Publishing.
Note: GM: geometric mean; FCs: fecal coliforms.

TABLE 6.4

Time-Temperature Matrix for *Vibrio parahaemolyticus*

Action Level	Average Monthly Maximum Air Temperature	Maximum Hours from Exposure to Temperature Control
Level 1	<65°F (18°C)	36 hours
Level 2	66°F–81°F (19°C–27°C)	12 hours
Level 3	≥81°F (≥27°C)	10 hours

national electronic norovirus outbreak surveillance network to facilitate investigation of foodborne norovirus outbreaks. It was launched in March 2009 and 20 states and local health laboratories had been certified to submit norovirus sequences and epidemiologic outbreak data to CaliciNet (Vega et al., 2011).

6.2.5.2 Depuration and Relaying

Depuration is a postharvest process where bivalve mollusks are immersed in tanks of clean seawater. Depuration consists of a flow-through or recirculation system of chemically (chlorine, ozone, iodophores, and activated oxygen) or physically (UV radiation) disinfected water to allow purification under controlled conditions (Lees, 2000; Richards, 1988). During depuration, fecal coliforms and other contaminants are expelled by the shellfish into the water, which is regularly exchanged (Teplitski et al., 2009). Just as they accumulate organisms when filtering contaminated water, these organisms may be removed when filtering clean water. Depuration periods may vary from 1 to 7 days, but a 2-day period is most widely used (Lees, 2000). Shellfish harvested from polluted areas can be transferred to cleaner estuaries to allow shellfish to cleanse or purge themselves. This process is called *relaying*. Shellfish can only be held for a short period in depuration tanks (i.e., 2 days), but can be held much longer in the natural environment. This makes relaying suitable to treat more heavily polluted shellfish where longer periods (i.e., a minimum of two months for Category C shellfish) are required to remove heavy contaminant loads (Lees, 2000). Purification processes are based on the assumption that if shellfish became contaminated by filtering polluted water, they can also purge the contaminants by filtering with clean water. Thus, microbial depuration and relaying decreases the risk for potential infections and were also less gritty and more palatable (Richards, 1988).

Phuvasate et al. (2012) demonstrated that holding oysters in ice or depuration of oysters in recirculating seawater at 36–37°F (2–3°C) for 4 days did not result in significant reductions of *V. parahaemolyticus*. The limited reduction is probably due to minimal biological activity of oysters at low temperatures. However, depuration at temperatures between 45–59°F (7–15°C) reduced *V. parahaemolyticus* populations in Pacific oysters by > 3.0 log MPN/g after 5 days. Depuration temperature at 59°F (15°C) was also reported to reduce *V. parahaemolyticus* and *V. vulnificus* in Gulf oysters by 2.1 and 2.9 log MPN/g (Chae et al., 2009). It seems that by reducing temperature during depuration it enhanced the reduction of *V. parahaemolyticus* in oysters. However, the temperature should be more than 4°F (5°C) since biological activity is minimal below this temperature (Su et al., 2010). Depuration of oysters at temperatures between 45–59°F (7–15°C) can be applied as a postharvest treatment (Phuvasate et al., 2012).

6.2.5.3 Good Manufacturing Practices and HACCP

In fish processing, the Hazard Analysis and Critical Control Point (HACCP) system is designed to identify hazards, establish critical control points (CCPs) and critical limits, monitor procedures, documentation, and recording. HACCP should be operated together with good manufacturing practices (GMP). Every seafood harvester and processor is required to use a HACCP-based system and identify hazards and critical control points from harvest to consumption (Amagliani et al., 2012). Some of the main postharvest CCPs for pathogen control in seafood include: primary chilling immediately in ice slurry on vessels and at harvest site; applying time–temperature regimens to give log reductions of contamination levels, rapid chilling after cooking, plate freezing, and frozen storage (Arvanitoyannis and Varsakas, 2009). Heat treatments reaching a core temperature above 140°F (60°C) and deep freezing below –4°F (–20°C) for at least 24 hours will kill all parasites commonly found in fish. Brining may reduce the parasite hazard, but since parasites are adapted to stomach conditions, low pH does not affect them significantly (Levsen and Lunestad, 2008). Finfish should be rapidly chilled or gutted after capture. A prolonged time before freezing may allow parasites to migrate from the gut to the flesh (Butt et al., 2004b).

An additional processing method such as high pressure (HP) processing has been investigated to ensure the safety of shellfish for human consumption. HP technology allows inactivation of microorganisms while maintaining sensory and nutritional properties of foods (Oliveira et al., 2011). Hepatitis A virus (HAV) (Kingsley et al., 2002) and HAV within contaminated shellfish was inactivated by HP processing (Calci et al., 2005). HP can be used to complement present sanitation guidelines to provide an additional measure of safety to shellfish designated for raw consumption and/or cooking (Calci et al., 2005). In addition to enhancing safety, HP treatment is also used in shucking and opening shellfish (Murchie et al., 2005) making this technology particularly beneficial to the shellfish processing industry.

References

Adak, G. L., S. M. Meakins, H. Yip, B. A. Lopman, and S. J. O'Brien. 2005. Disease risks from foods, England and Wales, 1996–2000. *Emerg. Infect. Dis.* 11: 365–372.

Afonso, C., H. M. Lourenço, A. Dias, M. L. Nunes, and M. Castro. 2007. Contaminant metals in black scabbard fish (*Aphanopus carbo*) caught off Madeira and the Azores. *Food Chem.* 101: 120–125.

Ahl, A. S., D. Nganwa, and S. Wilson. 2002. Public health considerations in human consumption of wild game. *Ann. N. Y. Acad. Sci.* 969: 48–50.

Altwegg, M., G. Martinetti Lucchino, J. Lüthy-Hottenstein, and M. Rohrbach. 1991. *Aeromonas*-associated gastroenteritis after consumption of contaminated shrimp. *Eur. J. Clin. Microbiol.* 10: 44–45.

Amagliani, G., G. Brandi, and G. F. Schiavano. 2012. Incidence and role of *Salmonella* in seafood safety. *Food Res. Int.* 45: 780–788.

Amiard, J.-C., C. Amiard-Triquet, L. Charbonnier, A. Mesnil, P. S. Rainbow, and W.-X. Wang. 2008. Bioaccessibility of essential and non-essential metals in commercial shellfish from Western Europe and Asia. *Food Chem. Toxicol.* 46: 2010–2022.

Andrews, L. S. 2004. Strategies to control *Vibrios* in molluscan shellfish. *Food Prot. Trends* 24: 70–76.

Appleyard, G. D. and A. A. Gajadhar. 2000. A review of trichinellosis in people and wildlife in Canada. *Can. J. Public Health* 91: 293–297.

Arvanitoyannis, I. S and T. H. Varzakas. 2009. Seafood. In *HACCP and ISO 22000: Application to Foods of Animal Origin,* ed. I. S. Arvanitoyannis, 377. Oxford: Blackwell Publishing.

Atanassova, V., J. Apelt, F. Reich, and G. Klein. 2008. Microbiological quality of freshly shot game in Germany. *Meat Sci.* 78: 414–419.

Australia Standard 4464:2007. 2007. Australian Standard for the Hygienic Production of Wild Game Meat for Human Consumption. FRSC Tehnical Report No. 2, 1–59. Victoria: CSIRO Publishing.

Autio, T., S. Hielm., M. Miettinen et al. 1999. Sources of *Listeria monocytogenes* contamination in a cold-smoked rainbow trout processing plant detected by pulsed-field gel electrophoresis typing. *Appl. Environ. Microbiol.* 65: 150–155.

Avagnina, A., D. Nucera, M. A. Grassi, E. Ferroglio, A. Dalmasso, and T. Civera. 2012. The microbiological conditions of carcasses from large game animals in Italy. *Meat Sci.* 91: 266–271.

Bandick, N. and A. Hensel. 2011. Zoonotic diseases and direct marketing of game meat: Aspects of consumer safety in Germany. In *Game Meat Hygiene in Focus: Microbiology, Epidemiology, Risk Analysis and Quality Assurance,* ed. P. Paulsen, A. Bauer, and M. Vodnansky, 93–100. Wageningen: Wageningen Academic Publishers.

Barrento, S., A. Marques, B. Teixeira, P. Vaz-Pires, M. L. Carvalho, and M. L. Nunes. 2008. Essential elements and contaminants in edible tissues of European and American lobsters. *Food Chem.* 111: 862–867.

Barrento, S., A. Marques, B. Teixeira, M. L. Carvalho, P. Vaz-Pires, and M. L. Nunes. 2009. Accumulation of elements (S, As, Br, Sr, Cd, Hg, Pb) in two populations of *Cancer pagurus*: Ecological implications to human consumption. *Food Chem. Toxicol.* 47: 150–156.

Bártová, E., K. Sedlák, and I. Literák. 2006. Prevalence of *Toxoplasma gondii* and *Neospora caninum* antibodies in wild boars in the Czech Republic. *Vet. Parasitol.* 142: 150–153.

Bateson, P. and E. L. Bradshaw. 1997. Physiological effects of hunting red deer (*Cervus elaphus*). *Proc. R. Soc. B* 264: 1707–1714.

Bensink, J. C., I. Ekaputra, and C. Taliotis. 1991. The isolation of *Salmonella* from kangaroos and feral pigs processed for human consumption. *Aust. Vet. J.* 68: 106–107.

Berg, D. E., M. A. Kohn, T. A. Farley, and L. M. McFarland. 2000. Multi-state outbreaks of acute gastroenteritis traced to fecal-contaminated oysters harvested in Louisiana. *J. Infect. Dis.* 181: S381–S386.

Berntssen, M. H. G., K. Julshamn, and A.-K. Lundebye. 2010. Chemical contaminants in aquafeeds and Atlantic salmon (*Salmo salar*) following the use of traditional— versus alternative feed ingredients. *Chemosphere* 78: 637–646.

Bradshaw, E. L. and P. Bateson. 2000. Welfare implications of culling red deer (*Cervus elaphus*). *Anim. Welf.* 9: 3–24.

Brands, D. A., A. E. Inman, C. P. Gerba et al. 2005. Prevalence of *Salmonella* spp. in oysters in the United States. *Appl. Environ. Microbiol.* 71: 893–897.

Brett, M. S. Y., P. Short, and J. McLauchlin. 1998. A small outbreak of listeriosis associated with smoked mussels. *Int. J. Food Microbiol.* 43: 223–229.

Burkhardt III, W. and K. R. Calci. 2000. Selective accumulation may account for shellfish-associated viral illness. *Appl. Environ. Microbiol.* 66: 1375–1378.

Butt, A. A., K. E. Aldridge, and C. V. Sanders. 2004a. Infections related to the ingestion of seafood Part I: Viral and bacterial infections. *Lancet Infect. Dis.* 4: 201–212.

Butt, A. A., K. E. Aldridge, and C. V. Sanders. 2004b. Infections related to the ingestion of seafood Part II: Parasitic infections and food safety. *Lancet Infect. Dis.* 4: 294–300.

Cabello, F. C., R. Espejo, M. C. Hernandez, M. L. Rioseco, J. Ulloa, and J. A. Vergaram. 2007. *Vibrio parahaemolyticus* O3:K6 epidemic diarrhea, Chile, 2005. *Emerg. Infect. Dis.* 13: 655–656.

Calci, K. R., G. K. Meade, R. C. Tezloff, and D. H. Kingsley. 2005. High-pressure inactivation of Hepatitis A virus within oysters. *Appl. Environ. Microbiol.* 71: 339–343.

Callahan, K. M., D. J. Taylor, and M. D. Sobsey. 1995. Comparative survival of hepatitis A virus, poliovirus and indicator viruses in geographically diverse seawaters. *Water Sci. Technol.* 31: 189–193.

Cartwright, K. A. V. and B. G. Evans. 1988. Salmon as a food-poisoning vehicle—two successive *Salmonella* outbreaks. *Epidemiol. Infect.* 101: 249–257.

Casoli, C., E. Duranti, F. Cambiotti, and P. Avellini. 2005. Wild ungulate slaughtering and meat inspection. *Vet. Res. Commun.* 29: 89–95.

Castro-Escarpulli, G., M. J. Figueras, G. Aguilera-Arreola et al. 2003. Characterisation of *Aeromonas* spp. isolated from frozen fish intended for human consumption in Mexico. *Int. J. Food Microbiol.* 84: 41–49.

CDC. 2005. Preventing lead poisoning in young children. A statement by the Centers for Disease Control and Prevention. http://www.cdc.gov/nceh/lead/publications/prevleadpoisoning.pdf (accessed March 23, 2012).

CDC. 2009. Cluster of ciguatera fish poisoning—North Carolina, 2007. *MMWR Morb. Mortal. Wkly. Rep.* 58: 283–285.

CDC. 2010. Human paragonimiasis after eating raw or undercooked crayfish—Missouri, July 2006–September 2010. *MMWR Morb. Mortal. Wkly. Rep.* 59: 1573–1576.

Chae, M. J., D. Cheney, and Y.-C. Su. 2009. Temperature effects on the depuration of *Vibrio parahaemolyticus* and *Vibrio vulnificus* from the American oyster (*Crassostrea viginica*). *J. Food Sci.* 74: M62–M66.

Chen, B.-Y., R. Pyla, T.-J. Kim, J. L. Silva, and Y.-S. Jung. 2010a. Prevalence and contamination patterns of *Listeria monocytogenes* in catfish processing environment and fresh fillets. *Food Microbiol.* 27: 645–652.

Chen, H.-C., Y.-R. Huang, H. H. Hsu et al. 2010b. Determination of histamine and biogenic amines in fish cubes (*Tetrapturus angustirostris*) implicated in a food-borne poisoning. *Food Control* 21: 13–18.

Chen, H.-C., H.-F. Kung, W. C. Chen et al. 2008. Determination of histamine and histamine-forming bacteria in tuna dumpling implicated in a food-borne poisoning. *Food Chem.* 106: 612–618.

Chou, C. L. 2007. A time series of mercury accumulation and improvement of dietary feed in net caged Atlantic salmon (*Salmo salar*). *Mar. Pollut. Bull.* 54: 720–725.

Clegg, F. G., S. N. Chiejina, A. L. Duncan, R. N. Kay, and C. Wray. 1983. Outbreaks of *Salmonella* Newport infection in dairy herds and their relationship to management and contamination of the environment. *Vet. Rec.* 112: 580–584.

Coburn, H., E. Snary, and M. Wooldridge. 2003. Hazards and Risks from Wild Game: A Qualitative Risk Assessment. Centre for Epidemiology and Risk Analysis, Veterinary Laboratories Agency, Weybridge. http://www.foodbase.org.uk//admintools/reportdocuments/660-1-1120_MO1025_Final_Report.pdf (accessed March 2, 2012).

Coburn, H. L., E. Snary, L. A. Kelly, and M. Wooldridge. 2005. Qualitative risk assessment of the hazards and risks from wild game. *Vet. Rec.* 10: 321–322.

Cody, S. H., M. K. Glynn, J. A. Farrar et al. 1999. An outbreak of *Escherichia coli* O157:H7 infection from unpasteurized commercial apple juice. *Ann. Intern. Med.* 130: 202–209.

Daskalov, H. 2006. The importance of *Aeromonas hydrophila* in food safety. *Food Control* 17: 474–483.

Dauphin, G., C. Ragimbeau, and P. Malle. 2001. Use of PFGE typing for tracing contamination with *Listeria monocytogenes* in three cold-smoked salmon processing plants. *Int. J. Food Microbiol.* 64: 51–61.

Doyle, A., D. Barataud, A. Gallay et al. 2004. Norovirus Foodborne Outbreaks Associated with Consumption of Oysters from the Etang De Thau, France, December 2002. *Eurosurveill.* 9: pii = 451. http://www.eurosurveillance.org/ViewArticle.aspx?ArticleId=451 (accessed December 30, 2009).

Drake, S. L., A. DePaola, and L.-A. Jaykus. 2000. An overview of *Vibrio vulnificus* and *Vibrio parahaemolyticus*. *Compr. Rev. Food Sci. Food Saf.* 6: 120–144.

Duan, J. and Y.-C. Su. 2005. Occurrence of *Vibrio parahaemolyticus* in two Oregon oyster-growing bays. *J. Food Sci.* 70: M58–M63.

Dunn, J. R., J. E. Keen, D. Moreland, and R. A. Thompson. 2004. Prevalence of *Escherichia coli* O157:H7 in white-tailed deer from Louisiana. *J. Wildl. Dis.* 40: 361–365.

Dworkin, M. S., H. R. Gamble, D. S. Zarlenga, and P. O. Tennican. 1996. Outbreak of trichinellosis associated with eating cougar jerky. *J. Infect. Dis.* 174: 663–666.

Easton, M. D. L., D. Luszniak, and E. Von der Geest. 2002. Preliminary examination of contaminant loadings in farmed salmon and commercial salmon feed. *Chemosphere* 46: 1053–1074.

EFSA. 2007. Public health risks involved in the human consumption of reptile meat: Scientific opinion of the panel on biological risks. *EFSA Journal* 578: 1–55.

Eglezos, S., B. Huang, and E. Stuttard. 2007. A survey of the microbiological quality of kangaroo carcasses processed for human consumption in two processing plants in Queensland, Australia. *J. Food Prot.* 70: 1249–1251.

El-Ghareeb, W. R., F. J. M. Smulders, A. M. A. Morshdy, R. Winkelmayer, and P. Paulsen. 2009. Microbiological condition and shelf life of meat from hunted game birds. *Eur. J. Wildl. Res.* 55: 317–323.

Ericsson, H., A. Eklöw, M. L. Danielsson-Tham et al. 1997. An outbreak of listeriosis suspected to have been caused by rainbow trout. *J. Clin. Microbiol.* 35: 2904–2907.

Feldhusen, F. 2000. The role of seafood in bacterial foodborne diseases. *Microbes Infect.* 2: 1651–1660.

Fensterheim, R. J. 1993. Documenting temporal trends of polychlorinated biphenyls in the environment. *Regul. Toxicol. Pharmacol.* 18: 181–201.

Field, R. A. 2004. Game. In *Encyclopedia of Meat Sciences*, ed. W. K. Jensen, C. Devinem, and M. Dikeman, 1302–1308. Amsterdam: Elsevier.

Fisher, P., L. C. Hoffman, and F. D. Mellett. 2000. Processing and nutritional characteristics of value added ostrich products. *Meat Sci.* 55: 251–254.

Foreyt, W. J., T. E. Besser, and S. M. Lonning. 2001. Mortality in captive elk from salmonellosis. *J. Wildl. Dis.* 37: 399–402.

Francis, S., J. Rowland, and K. Rattenbury. 1989. An outbreak of paratyphoid fever in the UK associated with a fish-and-chip shop. *Epidemiol. Infect.* 103: 445–448.

French, E., A., Rodriguez-Palacios, and J. T. LeJeune. 2010. Enteric bacterial pathogens with zoonotic potential isolated from farm-rasied deer. *Foodborne Pathog. Dis.* 7: 1031–1037.

Gaffuri, A., M. Giacometti, V. M. Tranquillo, S. Magnino, P. Cordioli, and P. Lanfranchi. 2006. Serosurvey of roe deer, chamois and domestic sheep in the Central Italian Alps. *J. Wildl. Dis.* 42: 685–690.

Gantzer, C., E. Dubois, J.-M. Crance. 1998. Influence of environmental factors on the survival of enteric viruses in seawater. *Oceanol. Acta* 21: 983–992.

García, E., L. Mora, P. Torres, M. I. Jercic, and R. Mercado. 2005. First record of human trichinosis in Chile associated with consumption of wild boar (*Sus scrofa*). *Memórias do Instituto Oswaldo Cruz*, vol. 100, No. 1, Rio de Janeiro.

Gaulin, C., I. Oicard, N. Cote, M. Huot, and J.-F. Proulx. 2000. Outbreak of trichinellosis in French hunters who ate Canadian black bear meat. *Can. Comm. Dis. Rep.* 32: 109–112.

Gauss, C. B. L., J. P. Dubey, D. Vidal et al. 2005. Seroprevalence of *Toxoplasma gondii* in wild pigs (*Sus scrofa*) from Spain. *Vet. Parasitol.* 131: 151–156.

Gauss, C. B. L., J. P. Dubey, D. Vidal et al. 2006. Prevalence of *Toxoplasma gondii* antibodies in red deer (*Cervus elaphus*) and other wild ruminants from Spain. *Vet. Parasitol.* 136: 193–2000.

Gessner, B. and J. Mclaughlin. 2008. Epidemiologic impact of toxic episodes: Neurotoxic toxins. In *Seafood and Freshwater Toxins: Pharmacology, Physiology, and Detection*, ed. L. M. Botana, 77–104. Boca Raton: CRC Press.

Gill, C. O. 2007. Microbiological conditions of meats from large game animals and birds. *Meat Sci.* 77: 149–160.

Gill, C. O., J. Bryant, and D. A. Brereton. 2000. The microbiological conditions of the carcasses of six species after dressing at a small abattoir. *Food Microbiol.* 17: 233–239.

Gobat, P.-F. and T. Jemmi. 1993. Distribution of mesophilic *Aeromonas* species in raw and ready-to-eat fish and meat products in Switzerland. *Int. J. Food Microbiol.* 20: 117–120.

Greenbloom, S. L., P. Martin-Smith, S. Isaacs et al. 1997. Outbreak of trichinosis in Ontario secondary to the ingestion of wild boar meat. *Can. J. Public Health* 88: 52–56.

Guillois-Becel, Y., E. Couturier, J. C. Le Saux et al. 2009. An oyster-associated hepatitis A outbreak in France in 2007. *Eurosurveill.* 14: pii = 19144. http://www.eurosur-veillance.org/ViewArticle.aspx?ArticleId=19144 (accessed December 31, 2009).

Haley, B. J., D. J. Cole, and E. K. Lipp. 2009. Distribution, diversity and seasonality of waterborne *Salmonellae* in a rural watershed. *Appl. Environ. Microbiol.* 75: 1248–1255.

Halliday, M. L., L.-Y. Kang, T.-K. Zhou et al. 1991. An epidemic of hepatitis A attributable to the ingestion of raw clams in Shanghai, China. *J. Infect. Dis.* 164: 852–859.

Håstein, T., B. Hjeltnes, A. Lillehaug, J. U. Skåre, M. Berntssen, and A. K. Lundebye. 2006. Food safety hazards that occur during the production stage: Challenges for fish farming and the fishing industry. *Sci. Technol. Rev.* 25: 607–625.

Hayashidani, H., N. Kanzaki, Y. Kaneko et al. 2002. Occurrence of yersiniosis and listeriosis in wild boars in Japan. *J. Wildl. Dis.* 38: 202–205.

Hayward, D., J. Wong, and A. J. Krynitsky. 2007. Polybrominated diphenyl ethers and polychlorinated biphenyls in commercially wild caught and farm-raised fish fillets in the United States. *Environ. Res.* 103: 46–54.

Heinitz, M. L., R. D. Ruble, D. E. Wagner, and S. R. Tatini. 2000. Incidence of *Salmonella* in fish and seafood. *J. Food Prot.* 63: 579–592.

Hercock, M. 2004. The wild kangaroo industry: Developing the potential for sustainability. *The Environmentalist* 24: 73–86.

Herreras, M. V., F. J. Aznar, J. A. Balbuena, and J. A. Raga. 2000. Anisakid larvae in the musculature of the Argentinean hake, *Merluccius hubbsi*. *J. Food Prot.* 63: 1141–1143.

Herrero, B., J. M. Vieites, and M. Espiñeira. 2011. Detection of anisakids in fish and seafood products by real-time PCR. *Food Control* 22: 933–939.

Hites, R. A., J. A. Foran, S. J. Schwager, B. A. Knuth, M. C. Hamilton, and D. O. Carpenter. 2004. Global assessment of polybrominated diphenyl ethers in farmed and wild salmon. *Environ. Sci. Technol.* 38: 4945–4949.

Hoffman, L. C., S. St. C. Botha, and T. J. Britz. 2006. Sensory properties of hot-deboned ostrich (*Struthio camelus* var. *domesticus*) *Mucularis gastrocnemius, pars interna*. *Meat Sci.* 72: 734–740.

Hoffman, L. C. and E. Wiklund. 2006. Game and venison—meat for the modern consumer. *Meat Sci.* 74: 197–208.

Holds, G., A. Pointon, M. Lorimer, A. Kiermeier, G. Raven, and J. Sumner. 2008. Microbial profiles of carcasses and minced meat from kangaroos processed in South Australia. *Int. J. Food Microbiol.* 123: 88–92.

Hunt, G., W. Burnham, C. Parish, K. Burnham, B. Mutch, and J. L. Oaks. 2006. Bullet fragments in deer remains: Implications for lead exposure in scavengers. *Wildlife Soc. B.* 34: 167–170.

Hunt, W. G., R. T. Watson, and J. L. Oaks. 2009. Lead bullet fragments in venison from rifle killed deer. *PloS ONE* 4: e5330.

Huss, H. H., A. Reilly, and P. K. B. Embarek. 2000. Prevention and control of hazards in seafood. *Food Control* 11: 149–156.

IDSC. 1999. Infectious Disease Surveillance Center. *Vibrio parahaemolyticus*, Japan, 1996–1998. *Infectious Agent Surveillance Report* 20: No.233. http://idsc.nih.go.jp/iasr/20/233/tpc233.html (accessed January 27, 2010).

Iqbal, S. 2008. Epi-Aid Trip Report: Assessment of Human Health Risk from Consumption of Wild Game Meat with Possible Lead Contamination among the Residents of the State of North Dakota. National Center for Environmental Health, Centers for Disease Control and Prevention, Atlanta, USA. http://www.rmef.org/NR/rdonlyres/F07627AA-4D94-4CBC-B8FD-4F4F18401303/0/ND_report.pdf (accessed March 23, 2012).

Iwamoto, M., T. Ayers, B. E. Mahon, and D. L. Swerdlow. 2010. Epidemiology of seafood-associated infections in the United States. *Clin. Microbiol. Rev.* 23: 399–411.

Johansson, T., L. Rantala, L. Palmu, and T. Honkanen-Buzalski. 1999. Occurrence and typing of *Listeria monocytogenes* strains in retail vacuum packed fish products and in a production plant. *Int. J. Food Microbiol.* 47: 111–119.

Johnson, L. L., G. M. Ylitalo, and C. A. Sloan. 2007. Persistent organic pollutants in outmigrant juvenile Chinook salmon from the Lower Columbia Estuary, USA. *Sci. Total Environ.* 374: 342–366.

Jones, S. H., T. L. Howell, and K. R. O'Neill. 1991. Differential elimination of indicator bacteria and pathogenic *Vibrio* spp. from eastern oysters (*Crassostrea virginica* Gmelin, 1971) in a commercial controlled purification facility in Maine. *J. Shellfish R.* 10: 105–112.

Jongwutiwes, S., N. Chantachum, P. Kraivichian et al. 1998. First outbreak of human trichinellosis caused by *Trichinella pseudospiralis*. *Clin. Infect. Dis.* 26: 111–115.

Jusko, T. A., C. R. Jr. Henderson, B. P. Lanphear, D. A. Cory-Slechta, P. J. Parsons, and R. L. Canfield. 2008. Blood lead concentrations < 10 μg/dL and child intelligence at 6 years of age. *Environ. Health Perspect.* 116: 243–248.

Kaneko, T. and R. R. Colwell. 1973. Ecology of *Vibrio parahaemolyticus* in Chesapeake Bay. *J. Bacteriol.* 113: 24–32.

Karunasagar, I. and I. Karunasagar. 2000. *Listeria* in tropical fish and fishery products. *Int. J. Food Microbiol.* 62: 177–181.

Keene, W. E., E. Sazie, J. Kok et al. 1997. An outbreak of *Escherichia coli* O157:H7 infections traced to jerky made from deer meat. *J. Am. Med. Assoc.* 277: 1229–1231.

Khumjui, C., P. Choomkasien, P. Dekumyoy et al. 2008. Outbreak of trichinellosis caused by *Trichinella papuae*, Thailand, 2006. *Emerg. Infect. Dis.* 14: 1913–1915.

Kingsley, D. H., D. Hoover, E. Papafragkou, and G. P. Richards. 2002. Inactivation of hepatitis A virus and a calicivirus by high hydrostatic pressure. *J. Food Prot.* 65: 1605–1609.

Leal, N. C., S. C. Da Silva, V. O. Cavalcanti et al. 2008. *Vibrio parahaemolyticus* serovar O3:K6 gastroenteritis in northeast Brazil. *J. Appl. Microbiol.* 105: 691–697.

Lee, W.-C., M.-J. Lee, J.-S. Kim, and S.-Y. Park. 2001. Foodborne illness outbreaks in Korea and Japan studied retrospectively. *J. Food Prot.* 64: 899–902.

Lees, D. 2000. Viruses and bivalve shellfish. *Int. J. Food Microbiol.* 59: 81–116.

Levsen, A. and B. T. Lunestad. 2008. Parasites in farmed fish and fishery products. In *Improving Farmed Fish Quality and Safety*, ed. L. Oyvind, 428–445. Cambridge: Woodhead Publishing Limited.

Lewis, R. J. 2001. The changing face of ciguatera. *Toxicon* 39: 97–106.

Ley, E. V., T. Y. Morishita, T. Brisker, and B. S. Harr. 2001. Prevalence of *Salmonella*, *Campylobacter*, and *Escherichia coli* on ostrich carcasses and the susceptibility of ostrich-origin *E. coli* isolates to various antibiotics. *Avian Dis.* 45: 696–700.

Lillehaug, A., B. Bergsjø, J. Schau, T. Bruheim, T. Vikøren, and K. Handeland. 2005. *Campylobacter* spp., *Salmonella* spp., verocytotoxic *Escherichia coli* and antibiotic resistance in indicator organisms in wild cervids. *Acta Vet. Scand.* 46: 23–32.

Liu, X., Y. Chen. X. Wang, and R. Ji. 2004. Foodborne disease outbreaks in China from 1992 to 2001 national foodborne disease surveillance system. *J. Hyg. Res.* 33: 725–727.

Lozano-León, A., J. Torres, C. R. Osorio, and J. Martinez-Urtaza. 2003. Identification of *tdh*-positive *Vibrio parahaemolyticus* from an outbreak associated with raw oyster consumption in Spain. *FEMS Microbiol. Lett.* 226: 281–284.

Lunestad, B. T. and J. T. Rosnes. 2008. Microbiological quality and safety of farmed fish. In *Improving farmed fish quality and safety*, ed. L. Oyvind, 399–427. Cambridge: Woodhead Publishing Limited.

Maage, A., H. Hove, and K. Julshamn. 2008. Monitoring and surveillance to improve farmed fish safety. In *Improving Farmed Fish Quality and Safety*, L. Øyvind. 547–564, Cambridge: Woodhead Publishing Limited.

Madsen, M. 1994. Enumeration of salmonellae in crocodile pond water by direct plate counts and by the mpn technique. *Water Res.* 28: 2035–2037.

Madsen, M. 1996. Prevalence and serovar distribution of *Salmonella* in fresh and frozen meat from captive Nile crocodiles (*Crocodylus niloticus*). *Int. J. Food Microbiol.* 29: 111–118.

Madsen, M., J. A. C. Milne, and P. Chambers. 1992. Critical control points in the slaughter and dressing of farmed crocodiles. *J. Food Sci. Technol.* 29: 265–267.

Magnino, S., P. Colin, E. Dei-Cas et al. 2009. Biological risks associated with consumption of reptile products. *Int. J. Food Microbiol.* 134: 163–175.

Maillard, D. and P. Fournier. 1995. Effects of shooting with hounds on size of resting range of wild boar (*Sus scrofa* L.) groups in Mediterranean habitat. *IBEX Journal of Mountain Ecology* 3: 102–107.

Malfait, P., P. L. Lopalco, S. Salmaso et al. 1996. An outbreak of Hepatitis A in Puglia, Italy, 1996. *Eurosurveill.* 1: pii = 144. http://www.eurosurveillance.org/ViewArticle.aspx?ArticleId=144 (accessed January 3, 2010).

Manchester-Neesvig, J. B., K. Valters, and W. C. Sonzogni. 2001. Comparison of polybrominated diphenyl ethers (PBDEs) and polychlorinated biphenyls (PCBs) in Lake Michigan salmonids. *Environ. Sci. Technol.* 35: 1072–1077.

Martinez-Urtaza, J., M. Saco, J. de Novoa et al. 2004. Influence of environmental factors and human activity on the presence of *Salmonella* serovars in a marine environment. *Appl. Environ. Microbiol.* 70: 2089–2097.

Martinez-Urtaza, J., M. Saco, G. Hernandez-Cordova, A. Lozano, O. Garcia-Martin, and J. Espinosa. 2003. Identification of *Salmonella* serovars isolated from live Molluscan shellfish and their significance in the marine environment. *J. Food Prot.* 66: 226–232.

Martinez-Urtaza, J., L. Simentall, D. Velasco et al. 2005. Pandemic *Vibrio parahemolyticus* O3:K6, Europe. *Emerg. Infect. Dis.* 11: 1319–1320.

Mateo, R., M. Rodríguez-de la Cruz, D. Vidal, M. Reglero, and P. Camarero. 2007. Transfer of lead from shot pellets to game meat during cooking. *Sci. Total Environ.* 372: 480–485.

Matsumoto, J., Y. Kako, Y. Morita et al. 2011. Seroprevalence of *Toxoplasma gondii* in wild boars (*Sus scrofa leucomystax*) and wild sika deer (*Cervus nippon*) in Gunma Prefecture, Japan. *Parasitol. Int.* 60: 331–332.

McCormick, R. J. 2003. Game—mammals. In *Encyclopedia of Food Sciences and Nutrition*, ed. B. Caballero, L. Trugo, and P. M., Finglas, 2856–2860. Amsterdam: Elsevier Science Ltd.

McIntyre, L., S. L. Pollock, M. Fyfe et al. 2007. Trichinellosis from consumption of wild game meat. *Can. Med. Assoc. J.* 176: 449–451.

McLaughlin, J. B., A. DePaola, C. A. Bopp et al. 2005. Outbreak of *Vibrio parahaemolyticus* gastroenteritis associated with Alaskan oysters. *N. Engl. J. Med.* 353: 463–470.

Membré, J.-M., M. Laroche, and C. Magras. 2011. Assessment of levels of bacterial contamination of large wild game meat in Europe. *Food Microbiol.* 28: 1072–1079.

Menke, A., P. Muntner, V. Batuman, E. K. Silbergeld, and E. Guallar. 2006. Blood lead level below 0.48 µmol/L (10 µg/and mortality among US adults. *Circulation* 114: 1388–1394.

Missildine, B. R., R. J. Peters, G. Chin-Leo, and D. Houck. 2005. Polychlorinated biphenyls in adult Chinook salmon (*Oncorhynchus tshawytscha*) returning to coastal and Puget Sound hatcheries of Washington state. *Environ. Sci. Technol.* 39: 6944–6951.

Montory, M., E. Habit, P. Fernandez, J. O. Grimalt, and R. Barra. 2010. PCBs and PBDEs in wild Chinook salmon (*Oncorhynchus tshawytscha*) in the Northern Patagonia, Chile. *Chemosphere* 78: 1193–1199.

Mozaffarian, D. and E. B. Rimm. 2006. Fish intake, contaminants and human health: Evaluating the risks and the benefits. *J. Am. Med. Assoc.* 296: 1885–1899.

Murchie, L. W., M. Cruz-Romero, J. P. Kerry et al. 2005. High pressure processing of shellfish: A review of microbiological and other quality aspects. *Innov. Food Sci. Emerg. Technol.* 6: 257–270.

Murrell, K. D. and E. Pozio. 2000. Trichinellosis: The zoonosis that won't go quietly. *Int. J. Parasitol.* 30: 1339–1349.

Nawa, Y., C. Hatz, and J. Blum. 2005. Sushi delights and parasites: The risk of fishborne and foodborne parasitic zoonoses in Asia. *Clin. Infect. Dis.* 41: 1297–1303.

Naya, Y., M. Horiuchi, N. Ishiguro, and M. Shinagawa. 2003. Bacteriological and genetic assessment of game meat from Japanese wild boars. *J. Agric. Food Chem.* 51: 345–349.

Noël, L., C. Chafey, C. Testu, J. Pinte, P. Velge, and T. Guérin. 2011. Contamination levels of lead, cadmium and mercury in imported and domestic lobsters and large crab species consumed in France: Differences between white and brown meat. *J. Food Compost. Anal.* 24: 368–375.

NSSP. 2009. Chapter VIII Control of Shellfish Harvesting. National Shellfish Sanitation Program Guide for the Control of Molluscan Shellfish 2009 revision. http://www.fda.gov/Food/FoodSafety/Product-SpecificInformation/Seafood/FederalStatePrograms/NationalShellfishSanitationProgram/ucm047104.htm (accessed March 6, 2012).

Ohta, S., D. Ishizuka, H. Nishimura et al. 2002. Comparison of ploybrominated diphenyl ethers in fish, vegetables, and meats and levels in human milk of nursing women in Japan. *Chemosphere* 46: 689–696.

Oliveira, J., A. Cunha, F. Castilho, J. L. Romalde, and M. J. Pereira. 2011. Microbial contamination and purification of bivalve shellfish: Crucial aspects in monitoring and future perspectives—A mini-review. *Food Control* 22: 805–816.

Ottaviani, D., F. Leoni, E. Rocchegiani et al. 2008. First clinical report of pandemic *Vibrio parahaemolyticus* O3:K6 infection in Italy. *J. Clin. Microbiol.* 46: 2144–2145.

Parameswaran, N., R. M. O'Handley, M. E. Grigg, S. G. Fenwick, and R. C. A. Thompson. 2009. Seroprevalence of *Toxoplasma gondii* in wild kangaroos using an ELISA. *Parasitol. Int.* 58: 161–165.

Paulsen, P. 2011. Hygiene and microbiology of meat from wild game: An Austrian view. In *Game Meat Hygiene in Focus: Microbiology, Epidemiology, Risk Analysis and Quality Assurance*, ed. P. Paulsen, A. Bauer and M. Vodnansky, 19–37. Wageningen: Wageningen Academic Publishers.

Paulsen, P., J. Nagy, P. Popelka et al. 2008. Influence of storage conditions and shotshell wounding on the hygienic condition of hunted, uneviscerated pheasant (*Phasianus colchicus*). *Poultry Sci.* 87: 191–195.

Paulsen, P. and R. Winkelmayer. 2004. Seasonal variation in the microbial contamination of game carcasses in an Austrian hunting area. *Eur. J. Wildl. Res.* 50: 157–159.

Phuvasate, S., M.-H. Chen, and Y.-C. Su. 2012. Reductions of *Vibrio parahaemolyticus* in Pacific oysters (*Crassostrea gigas*) by depuration at various temperatures. *Food Microbiol.* 31: 51–56.

Potasman, I., A. Paz, and M. Odeh, M. 2002. Infectious outbreaks associated with bivalve shellfish consumption: A worldwide perspective. *Clin. Infect. Dis.* 35: 921–928.

Potter, A. S., S. A. Reid, and S. G. Fenwick. 2011. Prevalence of *Salmonella* in fecal samples of western grey kangaroos (*Macropus fuliginosus*). *J. Wildl. Dis.* 47: 880–887.

Pozio, E. 2001. New patterns of *Trichinella* infection. *Vet. Parasitol.* 98: 133–148.

Puente, P., A. M. Anadón, M. Rodero, F. Romarís, F. M. Ubeira, and C. Cuéllar. 2008. *Anisakis simplex*: The high prevalence in Madrid (Spain) and its relation with fish consumption. *Exp. Parasitol.* 118: 271–274.

Rabatsky-Ehr, T., D. Dingman, R. Marcus, R. Howard, A. Kinney, and P. Mshar. 2002. Deer meat as the source for a sporadic case of *Escherichia coli* O157:H7 infection, Connecticut. *Emerg. Infect. Dis.* 8: 525–527.

Radu, S., N. Ahmad, H. L. Foo, and Abdul Reezal. 2003. Prevalence and resistance to antibiotics for *Aeromonas* species from retail fish in Malaysia. *Int. J. Food Microbiol.* 81: 261–266.

Ramanzin, M., A. Amici, C. Casoli et al. 2010. Meat from wild ungulates: Ensuring quality and hygiene of an increasing resource. *Ital. J. Anim. Sci.* 9: 318–331.

Regulation (EC) No. 852/2004 of the European Parliament and of the Council of April 29, 2004 on the hygiene of foodstuffs. *Off. J. Eur. Union L* 226/19. http://eur-lex.europa.eu/LexUriServ/LexUriServ.do?uri=OJ:L:2004:226:0003:0021:EN:PDF (accessed March 23, 2012).

Regulation (EC) No. 853/2004 of the European Parliament and of the Council of April 29, 2004 laying down specific hygiene rules for food of animal origin. *Off. J. Eur. Union L* 226/22. http://www.fsai.ie/uploadedFiles/Reg853_2004(1).pdf (accessed February 22, 2012).

Regulation (EC) No. 2075/2005. Commission Regulation EC 2075/2005 of December 2005 laying down specific rules on official controls for *Trichinella* in meat. http://eur-lex.europa.eu/LexUriServ/LexUriServ.do?uri=CELEX:32005R2075:en:NOT (accessed March 23, 2012).

Renter, D. G., J. M. Sargeant, S. E. Hygnstorm, J. D. Hoffman, and J. R. Gillespie. 2001. *Escherichia coli* O157:H7 in free-ranging deer in Nebraska. *J. Wildl. Dis.* 37: 755–760.

Rhodes, M. W. and H. Kator. 1988. Survival of *Escherichia coli* and *Salmonella* spp. in estuarine environments. *Appl. Environ. Microbiol.* 54: 2902–2907.

Riccardo, O., B. Cristina, P. Vittorio, and D. Lorenzo. 2002. First report of *Trichinella britovi* infection in the wild boar of Aosta Valley. *Z. Jagdwiss.* 48: 247–255.

Richards, G. P. 1988. Microbial purification of shellfish: A review of depuration and relaying. *J. Food Prot.* 51: 218–251.

Richards, P. J., S.Wu, D. B. Tinker, M. V. Howell, and C. E. R. Dodd. 2011. Microbial quality of venison meat at retail in the UK in relation to production practices and processes. In *Game Meat Hygiene in Focus: Microbiology, Epidemiology, Risk Analysis and Quality Assurance*, ed. P. Paulsen, A. Bauer, and M. Vodnansky, 113–117. Wageningen: Wageningen Academic Publishers.

Rørvik, L. M. 2000. *Listeria monocytogenes* in the smoked salmon industry. *Int. J. Food Microbiol.* 62: 183–190.

Sacks, J. J., D. G. Delgado, H. O. Lobel, and R. L. Parker. 1983. Toxoplasmosis infection associated with eating undercooked venison. *Am. J. Epidemiol.* 118: 1983.

Sales, J. 1998. Fatty acid composition and cholesterol content of different ostrich muscles. *Meat Sci.* 49: 489–492.

Sargeant, J. M., D. J. Hafer, J. R. Gillespie, R. D. Oberst, and S. J. Flood. 1999. Prevalence of *Escherichia coli* O157:H7 in white-tailed deer sharing rangeland with cattle. *J. Am. Vet. Med. Assoc.* 215: 792–794.

Scallan, E., R. M. Hoekstra, F. J. Angulo et al. 2011. Foodborne illness acquired in the United States—Major pathogens. *Emerg. Infect. Dis.* 17: 7–15.

Schellenberg, R. S., B. J. K. Tan, J. D. Irvine et al. 2003. An outbreak of trichinellosis due to consumption of bear meat infected with *Trichinella nativa*, in 2 Northern Saskatchewan communities. *J. Infect. Dis.* 188: 835–843.

Setti, I., A. Rodriguez-Castro, and M. P. Pata. 2009. Characteristics and dynamics of *Salmonella* contamination along the Coast of Agadir, Morocco. *Appl. Environ. Microbiol.* 75: 7700–7709.

Severini, M., D. Ranucci, D. Miraglia, and R. Branciari. 2003. Preliminary study on the microbiological quality of ostrich (*Struthio camelus*) carcasses dressed in small Italian abattoirs. *Ital. J. Food Sci.* 15: 295–300.

Simenthal, L. and J. Martinez-Urtaza. 2008. Climate patterns governing the presence and permanence of Salmonellae in coastal areas of Bahia de Todos Santos, Mexico. *Appl. Environ. Microbiol.* 74: 5918–5924.

Soriano, A., B. Cruz, L. Gómez, A. Mariscal, and A. García Ruiz. 2006. Proteolysis, physicochemical characteristics and free fatty acid composition of dry sausages made with deer (*Cervus elaphus*) or wild boar (*Sus scrofa*) meat: A preliminary study. *Food Chem.* 96: 173–184.

South African Ostrich Business Chamber. 2012. Ostrich products. http://www.ostrichsa.co.za/products.php (accessed March 2, 2011).

Stone, D. 2006. Polybrominated diphenyl ethers and polychlorinated biphenyls in different tissue types from Chinook salmon (*Oncorhynchus tshawytscha*). *Bull. Environ. Contam. Toxicol.* 76: 148–154.

Storelli, M. M., G. Normanno, G. Barone et al. 2012. Toxic metals (Hg, Cd, and Pb) in fishery products imported into Italy: Suitability for human consumption. *J. Food Prot.* 75: 189–194.

Su, Y.-C. and C. Liu. 2007. *Vibrio parahaemolyticus*: A concern of seafood safety. *Food Microbiol.* 24: 549–558.

Su, Y.-C., Q. Yang, and C. Häse. 2010. Refrigerated seawater depuration for reducing *Vibrio parahaemolyticus* contamination in Pacific oyster (*Crassostrea gigas*). *J. Food Prot.* 73: 1111–1115.

Svendsen, T. C., K. Vorkamp, B. Rønsholdt, and J.-O. Frier. 2007. Organochlorines and polybrominated diphenyl ethers in four geographically separated populations of Atlantic salmon (*Salmo salar*). *J. Environ. Monit.* 9: 1213–1219.

Szlinder-Richert, J., I. Barska, Z. Usydus, and R. Grabic. 2010. Polybrominated diphenyl ethers (PBDEs) in selected fish species from the southern Baltic Sea. *Chemosphere* 78: 695–700.

Taggart, M. A., M. M. Reglero, P. R. Camarero, and R. Mateo. 2011. Should legislation regarding maximum Pb and Cd levels in human food also cover large game meat? *Environ. Int.* 37: 18–25.

Taylor, S. L., J. E. Stratton, and J. A. Nordlee. 1989. Histamine poisoning (scombroid fish poisoning): An allergy-like intoxication. *Clin. Toxicol.* 27: 225–240.

Teplitski, M., A. C. Wright, and G. Lorca. 2009. Biological approaches for controlling shellfish-associated pathogens. *Curr. Opin. Biotechnol.* 20: 185–190.

Tsai, Y. H., H.-F. Kung, T. M. Lee et al. 2005. Determination of histamine in canned mackerel implicated in a foodborne poisoning. *Food Control* 16: 579–585.

Urquhart, K. A. and I. J. McKendrick. 2006. Prevalence of "head shooting" and the characteristics of the wounds in culled wild Scottish red deer. *Vet. Rec.* 159: 75–79.

Vega, E., L. Barclay, N. Gregoricus, K. Williams, D. Lee, and J. Vinjé. 2011. Novel surveillance network for norovirus gastroenteritis outbreaks, United States. *Emerg. Infect. Dis.* 17: 1389–1395.

Vikøren, T., J. Tharaldsen, B. Fredriksen, and K. Handeland. 2004. Prevalence of *Toxoplasma gondii* antibodies in wild red deer, roe deer, moose, and reindeer from Norway. *Vet. Parasitol.* 120: 159–169.

Vivekanandhan, G., A. A. M. M. Hatha, and P. Lakshmanaperumalsamy. 2005. Prevalence of *Aeromonas hydrophila* in fish and prawns from the seafood market of Coimbatore, South India. *Food Microbiol.* 22: 133–137.

Wacheck, S., M. Fredriksson-Ahomaa, M. König, A. Stolle, and R. Stephan. 2010. Wild boars as an important reservoir for foodborne pathogens. *Foodborne Pathog. Dis.* 7: 307–312.

Wahlström, H., E. Tysén, E. O. Engvall et al. 2003. Survey of *Campylobacter* species, VTEC O157 and *Salmonella* species in Swedish wildlife. *Vet. Rec.* 153: 74–80.

Wallace, B. J., J. J. Guzewich, M. Cambridge, S. Altekruse, and D. L. Morse. 1999. Seafood-associated disease outbreaks in New York, 1980–1994. *Am. J. Prev. Med.* 17: 48–54.

WHO. 2010. *Safe Management of Shellfish and Harvest Waters*, ed. G. Rees, K. Pond, D. Kay, J. Bartram, and J. Santo Domingo. London: IWA Publishing.

Wilson, C. J. 2005. Feral Wild Boar in England: Status, Impact and Management. A report on behalf of Defra European Wildlife Division.

Zhang, L. and M. H. Wong. 2007. Environmental mercury contamination in China: Sources and impacts. *Environ. Int.* 33: 108–121.

7

Managing Food Safety Risks in Farmed Fish and Shellfish

7.1 Managing Food Safety Risks in Farmed Fish

7.1.1 Introduction

Historically, the oceans were considered limitless and thought to harbor enough fish to feed the world population. Now, aquaculture production—fish and shellfish farming—has grown rapidly to address the decreasing level of wild fisheries (Lunestad and Rosnes, 2008). Aquaculture is the farming of aquatic organisms, including finfish and shellfish, by individuals, groups, or corporations using some form of intervention in the rearing process (e.g., feeding, medications, regular stocking, protection from predators) that enhance production (FAO, 2009a). Although capture fisheries dominate world output, aquaculture accounts for a growing percentage of total fish supply, rising from a share of 4% in 1970 to 38% in 2009 (FAO, 2012). For example, the Atlantic salmon (*Salmo salar*) is the most significant aquaculture species in Europe, both in terms of production and economic value, with Norway, followed by Scotland and Ireland as the three major European producers (FAO, 2006). Norway and Chile are the world's leading producers of farmed salmon, accounting for 33% and 31% of world production, respectively (FAO, 2009b), followed by the United Kingdom and Canada. These four countries together produce over 85% of the world's farmed salmon (Liu and Sumaila, 2008). Farmed salmon production has increased from around 500 tons in 1970 (FISHSTAT, 1998) to 2.3 million tons in 2007 (FISHSTAT, 2007). In 1999, world farm salmon production for the first time surpassed salmon fishery production (Eagle et al., 2004).

The increase in seafood production through aquaculture provides a good source of high-quality protein and is an important cash crop in many parts of the world (Sapkota et al., 2008). The global aquaculture production is dominated by Asia (particularly China) which represents 11 of the top 15 aquaculture producing countries and accounts for 94% of the total global production (FAO, 2009c). The United States consumes 24.05 kg of fish/capita/year while the European Union consumes 22.03 kg fish/capita/year. Japan has the highest global per capita levels of fish consumption (60.78 kg/capita/year) while China consumes 26.46 kg/capita/year (Amagliani et al., 2012). The hazards for raw aquaculture products are different than those found in raw seafood caught in the open ocean. The two main hazards associated with (usually imported) aquaculture products are residues of unapproved drugs and contamination from foodborne pathogens during processing (Koonse, 2006).

7.1.2 Pathogenic Bacteria

The majority of human pathogens associated with aquaculture products are to be found in freshwater farms, farms in tropical countries, and among shellfish operations (Fairgrieve and Rust, 2003). Specific hazards to human health that might be associated with net-pen farming of salmon are anisakiasis, diphylobothriasis, rickettsialosis, vibriosis, aeromonasis, salmonellosis, and plesiomonasis (Fairgrieve and Rust, 2003). There have been very little reported cases of these hazards associated with farmed salmon. WHO (1999) also reported that the risk of contracting these illnesses from farmed fish is considered to be low. This is in agreement with Huss et al. (2000), who also reported that the level of human pathogenic bacteria in fish is generally quite low. The prevalence of *L. monocytogenes* range from 0 to 1% (Autio et al., 1999; Johansson et al., 1999) but a higher prevalence may occur from fish obtained from areas receiving heavy run-off from land (Huss et al., 1995). For example, Miettinen and Wirtanen (2006) found that weather conditions played an important role in the occurrence of *Listeria* spp. in a fish farm environment that was located near the shore. There was more *Listeria* spp. at the farm during rainy periods than during dry periods. Brook, river, and other run-off waters seemed to be the main routes for spreading *Listeria* spp. contamination to the farm. Pulsed field gel electrophoresis (PFGE) showed that *L. monocytogenes* isolates from the final fish products were similar to both isolates obtained from fish processing factories and from raw fish. This emphasizes that in addition to fish processing factory environment, the fish raw materials are important sources of *L. monocytogenes* contamination in final products.

Salmon products, especially cold-smoked salmon has always been associated with *Listeria monocytogenes* (Beaufort et al., 2007; Jørgensen and Huss, 1998; Lappi et al., 2004; Midelet-Bourdin et al., 2010). In contrast, *Salmonella* spp. and *Anisakis* spp. were not considered important hazards at the farm level. This is in agreement with Lunestad et al. (2007) who investigated the prevalence of *Salmonella* in samples of fish feed, feed ingredients, and environmental samples from feed producers in Norway. It was found that the prevalence of *Salmonella* in ready-to-use compound fish feed was 0.3%, and the most common serovars were S. Seftenberg, S. Agona, S. Montevideo, and S. Kentucky. The prevalence in feed ingredients varied from 0.14–0.33% and the prevalence in environmental samples was found to be 3.78%. *Salmonella* in feed materials may infect terrestrial animals (Lunestad et al. 2007), but *Salmonella* has not been regarded as a fish pathogen, with the possible exception of *Salmonella arizonae* (Kodama et al., 1987). Under natural rearing conditions of farmed salmon and low concentrations of *Salmonella* in feed, the risk of *Salmonella* in fish feed being transmitted to consumers are minimal (Lunestad et al., 2007; Nesse et al., 2005).

In terms of parasitic hazard, Marty (2008) found that 1 in 894 farmed salmon samples from Canada were infected with *Anisakis* spp. Nevertheless, he reported that the risk ratio of anisakid parasites in farmed salmon is 570 times less than in wild Atlantic salmon. The results were consistent with a previous study that was based on 0 anisakids in 3699 farmed salmon from Norway and Scotland (Angot and Brasseur, 1993). (This freedom from nematodes is recognized by the European Commission and farmed salmon were excused from the provisions of the hygiene regulations that require minimally processed fish intended for consumption without cooking to be frozen before sale). Presumably, this is because farmed salmon are fed processed feed and do not consume infected natural food. However, it must not be assumed that farmed salmon are free from infection, as in some systems, the cultured fish were fed with raw fish and the farmed product can become infected (Howgate, 1998).

7.1.3 Chemical Contaminants

7.1.3.1 Environmental Contaminants

Hites et al. (2004a) conducted a global assessment to determine the level of contaminants between farmed and wild salmon. The researchers revealed that the contaminants were significantly higher in farmed salmon from Europe than the farmed salmon from North America and wild salmon. Polychlorinated biphenyls (PCBs), dioxin, toxaphene, and dieldrin concentrations were also highest in farmed salmon from Scotland and the Faroe Islands. Other examples of organochlorine in farmed fish are polybrominated diphenyl ethers (PBDEs) in the United States and Canada (Hayward et al., 2007; Shaw et al., 2008), PBDEs in farmed salmons from Norway (Shaw et al., 2008), and PBDEs in farmed salmons from Europe (Hites et al., 2004b). According to Darnerud et al. (2006) and Kiviranta et al. (2004), fish is believed to be the main contributor of PBDEs in diet. The general observation is that the level of organic contaminants is higher in farmed salmon compared to wild salmon (Hayward et al., 2007; Hites et al., 2004a; Shaw et al., 2008). The consumption of these farmed fish can thus raise risk to health consequences such as cancer (Foran et al., 2005; Hites et al., 2004a, b). However, this is contradictory with Dewailly et al. (2007) who indicated that the overall concentrations of persistent organic pollutants (POPs) were low in both farmed and wild salmon; hence, the regular consumption of these fish would not cause significant health risks.

Polychlorinated dibenzodioxins (PCDDs) and polychlorinated dibenzo-furans (PCDFs) (dioxins) can be produced naturally due to forest fires or incomplete combustion of organic matter and industrial chemical processes. Dioxins are produced when mixtures of hydrocarbons and chlorine are exposed to high temperatures and are widely dispersed in the environment and are present particularly in lipid-rich food such as milk, meat, and fish. Dioxin exposure can result in liver damage, negative effects on immune and nervous systems (Berntssen and Lundebye, 2008). Similar to dioxins, poly-chlorinated biphenyls (PCBs) are ubiquitous in environment and are very persistent and lipid soluble. Some PCBs are very similar in structure to diox-ins and exert dioxin-like biological effects. PCBs arise from man-made prod-ucts such as electrical transformers, heat exchange fluids, hydraulic oils, and plastic manufacture. Although the production of PCBs is now banned, the chemicals have been deposited in the oceans (Bell et al., 2005) and are bioac-cumulated in the food chain. The maximum permitted level for dioxins alone and the sum of dioxins and dioxin-like PCBs in muscle meat from fish for human consumption are 4 and 8 pg WHO-TEQ/g (WHO-Toxic Equivalent) fresh weight, respectively (EC, 2006), with the exception of muscle meat from eel for which the maximum levels are 4 and 12 pg WHO-TEQ/g fresh weight. PCBs in farmed salmon vary considerably in farmed salmon with concentra-tions at 9 ng/g wwt (wet weight) to 51 ng/g wwt. Feed also varies from as low as 0.6 ng/g lipid to 1153 ng/g lipid (Table 7.1).

PBDEs in farmed fish such as rainbow trout measured less than 0.1 ng/g wwt (Montory et al., 2012) to about 4 ng/g wwt in farmed salmon (Usydus et al., 2009) (Table 7.1). A study performed by Schecter et al. (2004) also found that salmon fillets possess the highest total of PBDE concentrations, reaching 3078 parts per trillion (ppt) or pg g^{-1} (wet weight)—higher than the values found in meat and dairy products. PBDEs are brominated flame retardants that are industrially produced to reduce the risk and spread of fire. It is com-monly used in items such as television, computers, cars, and construction materials. PBDE is not considered genotoxic or carcinogenic to humans (EU, 2000, 2002, 2003) but may disrupt thyroid hormones. Currently, no inter-national legislation exists for safe levels of PBDEs in food (Berntssen and Lundebye, 2008).

The farmed salmon higher level of organochlorine could be due to the elevated level of contamination found in commercial salmon feed (Berntssen et al., 2010; Easton et al., 2002). Furthermore, the lipophilic nature of the con-taminants are likely to persist in fish oils and meals used in the production of salmon feeds (Chou, 2007). Once consumed, POPs have a strong tendency to bioaccumulate in the fatty tissues of fishes (Blanco et al., 2007). Fish oil is the main source of POPs while fish meal is the main source of mercury, cadmium, and arsenic (Berntssen et al., 2010). Another source of contamina-tion could be due to the inappropriate location of aquaculture sites with high natural contaminations (Jang et al., 2006). The differences in concentration of environmental contaminants are also due to the differences in feeding

TABLE 7.1

Chemical Contaminants Found in Farmed Salmon and Feed Samples

Samples	PCDD/Fs	PCBs	PBDEs	DDT	Endosulfan	References
			Contaminants			
Coho salmon Rainbow trout Salmon (Norway)			0.18 ng/g wwt 0.096 ng/g wwt			Montory et al., 2012
Fish feed (Norway)	0.34 pg g⁻¹ wwt	11 ng/g (Based on PCB₇) 13 ng/g (Based on PCB₇)	1.3 ng/g			NIFES, 2010 Maage et al., 2008
Salmon (Poland)		42.65 µg/kg wwt (Based on PCB₇)	3.96 ng/g wwt	50.0 µg kg⁻¹ wwt		Usydus et al., 2009
Farmed salmon		9.0 ng/g wwt	1.1 ng/g wwt			Hayward et al., 2007
Farmed salmon Feed (EU)		145–460 ng/g lipid 76–1153 ng/g lipid	1–85 ng/g lipid 8–24 ng/g lipid	5–250 ng g⁻¹ lipid 34–52 ng g⁻¹ lipid		Jacobs et al., 2002a
Farmed salmon		51.22 ng/g wwt	2.67 ng/g wwt			Easton et al., 2002
Salmon feed					0.058 ng/g wwt	
Fish feed (US)	1.76–38.08 pg g⁻¹ wwt	1.94 ng/g wwt		11.33 µg kg⁻¹ wwt		Maule et al., 2007
Feed	77 ± 50 pg g⁻¹ lipid	0.6 ng/g lipid		270 ng g⁻¹ lipid		Kelly et al., 2008
Farmed salmon	0.50 pg g⁻¹ wwt	21 ng/g wwt				Mozaffarian and Rimm, 2006

Continued

TABLE 7.1 (*Continued*)

Chemical Contaminants Found in Farmed Salmon and Feed Samples

Samples	Contaminants					References
	PCDD/Fs	PCBs	PBDEs	DDT	Endosulfan	
Salmon (Spain; purchased from markets)	7.33 pg g^{-1} wwt	12.16 ng/g wwt				Bocio et al., 2007
Salmon (US)		370.7 ng/g wwt		169.87 ng g^{-1} wwt		O'Toole et al., 2006
Farmed salmon (Canada)		12.9 ng/g wwt				Rawn et al., 2006
Farmed salmon			0.4–1.4 ng/g wwt			Shaw et al., 2008
Farmed salmon (Chile)			1.46 ng/g wwt			Montory and Barra, 2006
Feed			0.21–5.4 ng/g wwt			
Farmed salmon			0.97 ng/g wwt			Ohta et al., 2002
Salmon (Belgium)			1.58 ng/g wwt			Voorspoels et al., 2007
Salmon (Poland)				75–230 ng g^{-1} lipid		Szlinder-Richert et al., 2008

habits. For example, salmon caught in coastal waters of highly urbanized areas may exhibit higher PBDE levels (Hites et al., 2004b).

Heavy metals such as mercury, arsenic, lead, and cadmium were found in farmed salmon and feed in European Union but all were considerably below the maximum permitted level (Table 7.2). The low heavy metal concentrations may be due to farmed salmon being fed pellets and are usually harvested before 3 years age; hence, there is less opportunity for heavy metals to accumulate. In a separate study, Yamashita et al. (2005) evaluated mercury concentrations in commercially available fishes in Japan and found that farm-raised blue fin tuna had higher concentrations of mercury and methyl mercury compared to wild blue fin tuna caught from the same region. It was suggested that the use of large predatory fish such as mackerel as feed for the farmed tuna might have contributed to the higher levels of mercury observed in the fishes. This is, however, contradictory to Alam et al. (2002) studies who determined that there were no significant differences in metal concentrations between wild and cultured fish originated from the same polluted lake in Japan. This suggests that pathways for heavy metal uptake, target organs, and organism sensitivity are highly variable and are dependent on factors such as metal concentrations, age, size, physiological status, habitat preferences, feeding behavior, and growth rates of fish (Chapman et al., 1996).

7.1.3.2 Veterinary Residues and Antibiotics

In the European Union, United States, Canada, and Norway, there are strong regulatory control and limited use of antibiotics because of good management and development of effective vaccines (Lillehaug et al., 2003; WHO, 2002). In Australia and New Zealand, rigorous regulatory control is also adopted and no antibiotics are registered for use in aquaculture. However, ninety percent of the total world aquaculture production occurs in developing countries where regulatory controls are weak and uses of antibiotics are widespread (Bondad-Reantaso et al., 2005). This will be discussed in Section 8.7 under shrimp production. Studies by Akinbowale et al. (2006) suggested that there has been significant off-label use, when the researchers discovered that a number of bacteria isolated from aquaculture species were resistant towards a number of antibiotics.

7.1.3.3 Pesticides

Fish can be exposed to pesticides, their residues and metabolites from a variety of sources (Little et al., 2008). Pesticides may be applied intentionally to the fish to treat diseases, for example, the application of Cypermethrin bath treatments to control sea lice in salmon (Bodensteiner et al., 2000). Another route involves the inclusion of pesticides into the fish's diet intentionally, for example, in feed treatments containing emamectin benzoate (an ectoparasite

TABLE 7.2

Heavy Metal Concentration Data (mg/kg)

	Contaminants								
Matrix	Mercury (mg/kg wwt)	Maximum Permitted Level in EU	Arsenic, Total (mg/kg)	Maximum Permitted Level in EU	Cadmium (mg/kg)	Maximum Permitted Level in EU	Lead (mg/kg)	Maximum Permitted Level in EU	References
Salmon (Norway)	0.03	0.5 mg/kg wwt for most species; 0.5 mg/kg in marine feed ingredients; 0.1 mg/kg in feed (88% dry matter basis)	1.3	No EU maximum level for arsenic in food, but upper limits were set: 6 mg/kg in fish feed and 15 mg/kg in marine feed ingredients (88% dry matter)	<0.003	0.05 mg/kg wwt in most fish species	<0.02	0.2 mg/kg wet body weight for most fish species	Berntssen and Lundebye, 2008; NIFES, 2010
Fish feed (Norway)	0.07		5.4		0.33		0.11		NIFES, 2010
Salmon (Poland)	71.0 µg kg^{-1} wwt		1135 µg kg^{-1} wwt		8.0 µg kg^{-1} wwt		2.0 µg kg^{-1} wwt		Usydus et al., 2009
Farmed salmon (UK)	0.103								Knowles et al., 2003
Salmon (Canada)	0.05								Legrand et al., 2005
Salmon (Norway)	0.086								
Salmon (Alaska)	0.117								Plessi et al., 2001
Fish feed (Norway)			3.4–8.3						Sloth et al., 2005

agent), or unintentionally, for example, through exposure to residues such as endosulfan in the raw materials of formulated feed (Glover et al., 2007). Partial replacement of fish meal and oil with terrestrial, plant-based ingredients are thought to give a nutritionally adequate alternative as fish feed (Espe et al., 2007; Stubhaug et al., 2005) and can lower the content of environmental contaminants such as PCBs and dioxins (Berntssen et al., 2010).

However, the use of plant-based feed ingredients potentially exposes farmed fish to pesticides. One example is the organochloride endosulfan, which has been widely used as a broad-spectrum insecticide for the control of numerous insects in a wide variety of food and nongood crops (Sohn et al., 2004), and endosulfan poisoning is among the most frequently reported causes of aquatic organism incidents for pesticides (Dorval et al., 2003). Endosulfan has been reported in salmon feed (58 pg/g) (Easton et al., 2002). There is a possibility of pesticides carryover from feed to farmed fish (e.g., endosulfan) due to the increasing use of plant oil to substitute fish. Endosulfan were measured at 90 µg kg^{-1} in olive oil, which if incorporated into fish feeds, would exceed the European regulatory limit of 0.05 µg/kg (Directive 2002/32/EC 2002). The dietary transfer of endosulfan from feed to fish fillet is relatively low for Atlantic salmon, but feed concentrations that greatly exceed the current limit for endosulfan in fish feed (5 ppb) could cause a risk for food safety (Berntssen et al., 2008). Dichlorodiphenyltrichloroethane (DDT) on the other hand is not a significant hazard due to the declining presence of DDT in the environment and hence its deteriorating bioaccumulation in the food chain. Although its use has been phased out, traces of DDT can still be found in fish (Table 7.1).

7.1.3.4 Feed

Studies suggested that feed based on marine fish oils are likely to contaminate farmed salmon with PCBs and dioxins (Jacobs et al., 2002b). Bell et al. (2005) suggested that salmon fed on diets based on fish meal and oil attain flesh dioxin values ranging from 0.16–1.40 ng TEQ kg^{-1} and dioxin-like polychlorinated biphenyl (DL-PCB) concentrations ranging from 0.62–3.68 ng TEQ kg^{-1}. SCAN (2000) indicated that virtually most of basic feed ingredients, including fish meal and fish oil, are contaminated with dioxins at varying degrees. Feedstuffs originating from plants generally contain low levels of dioxins (0.1–0.2 ng kg^{-1} dry wt), while fish meal and oil, especially those originating from European sources are highly contaminated (fish meal 1.2 ng kg^{-1} dry wt; fish oil 4.8 ng kg^{-1} dry wt).

Salmon feed can contain up to 30% fish oil and 50% fish meal (Horst et al., 1998). Fish oil is a by-product of the fish meal manufacturing industry and is considered the main source of persistent organic pollutants in farmed fish (Easton et al., 2002; Jacobs et al., 2002a; WHO, 1999). Highly chlorinated compounds are lipophilic and bioaccumulate in the food chain. As they accumulate in the lipid compartment of the animal, the fish oil extracted from fish caught in polluted waters may be contaminated with chlorinated contaminants (Jacobs et al., 2002a). The researchers suggested that by replacing marine fish oils with vegetable oils, dioxin, and DL-PCB concentrations can be substantially reduced (Bell et al., 2005).

7.2 Managing Food Safety Risks in Shrimp

Aquaculture crustaceans are decapods such as prawns or shrimps, crabs, and lobsters (Figure 7.1). Decapod production is dominated by the marine penaied shrimps with the black tiger prawn (*Penaeus monodon*) being the most widely cultivated crustacean worldwide (Duncan, 2003). The terms "shrimp" and "prawn" are interchangeably used for different species in different parts of the world. However, the FAO convention is to call the marine and brackishwater forms "shrimps" and the freshwater forms "prawns" (FAO, 2008). Asia is by far the top producers for crustacean products with China producing more than 5 million metric tons in 2008, followed by Indonesia (772,000 metric tons), Thailand (647,000 metric tons), Vietnam (572,000 metric tons), and Canada (334,000 metric tons; FAO, 2010). A total of 3495 thousand metric tons of shrimps ($14 million) were produced in 2009 compared to 1100 thousand metric tons ($7 million) in 2000 (FAO, 2009c). The intensive aquaculture methods practiced may result in high stocking densities, high farm densities

FIGURE 7.1
Shrimp harvesting (from a shrimp farm) in Malaysia.

in coastal waters, and lack of appropriate barriers between farms; hence, the risk of bacterial infections is high. Diseases are known to curtail shrimp production by up to 70% (Nawaz et al., 2012). Therefore, heavy amounts of antibiotics are administered in feed for prophylactic (disease prevention) and therapeutic (disease treatment) purposes (Gräslund and Bengtsson, 2001; Holmström et al., 2003; Bondad-Reantaso et al., 2005).

7.2.1 Food Safety Hazards in Shrimp Production

7.2.1.1 Microbiological Hazards

Salmonella and *Listeria* are examples of pathogens associated with shrimp. These pathogens have been isolated from shrimp farms (Bhaskar and Sachinda, 2006), processing plants (Gudmundsdóttir et al., 2006; Vogel et al., 2001), and at retails (Phan et al., 2005). There have been at least 155 crustacean-associated outbreaks resulting in 1965 cases from 1998 to 2004 (NAMCF, 2008). *Salmonella* was found in pond sediment and water of a semi-intensive shrimp farm in India (Bhaskar et al., 1995). The major source of *Salmonella* in sediment is due to untreated animal manure used in the shrimp farm for purposes of pond fertilization. Similar to finfish, shrimp are also at risk from contamination in the processing plant from the processing environment, water used in processing, equipment, and workers (Valdimarsson et al., 1998). Shrimp picked by hand (especially during sorting and grading) can be easily contaminated with foodborne pathogens through poor personal hygiene practices.

Vibrio harveyi and *V. anguillarum* have been detected in shrimp hatcheries (Vaseeharan and Ramasamy, 2003), *V. alginolyticus*, *V. parahaemolyticus*, *V. harveyi*, and *V. vunificus* were detected in shrimp farms in India (Gopal et al., 2005) and *V. cholerae* O1 and *V. cholerae* non-O1 were isolated from 2% and 33% of samples collected from a major shrimp producing area in Southern Thailand. The results indicated that *V. cholerae* non-O1 is ubiquitous in aquatic environments where shrimp culture is practiced. Live giant tiger prawns obtained from markets in China were also found to be heavily contaminated with *V. vulnificus* (Yano et al., 2004) and *V. parahaemolyticus* (Yano et al., 2006) and *V. parahaemolyticus* were detected in frozen ready-to-eat shrimps in Malaysia (Sujeewa et al., 2009). The most important *Vibrio* species associated with disease problems in humans due to ingestion or other routes of exposure are *V. cholerae*, *V. parahemolyticus,* and *V. vulnificus* (FAO, 2001; Gopal et al., 2005).

V. parahaemolyticus has resulted in outbreaks in Mexico (in 2003 and late September of 2004) where more than 1230 cases of gastroenteritis were reported. All cases were attributed to consumption of raw or undercooked shrimp harvested from a lagoon (Cabanillas-Beltrán et al., 2006). The environmental conditions of the lagoon in September and October experienced elevated temperatures of >86°F (>30°C) and low salinities (<10 ppt) and were appropriate for the proliferation of this pathogen (DePaola et al., 2003).

7.2.1.2 Antibiotics and Antibiotic-Resistant Bacteria

It is well known that antibiotics are used in aquaculture, especially in shrimp farming to prevent or treat disease outbreaks. Studies conducted by Gräslund et al. (2003) revealed that 74% of the 76 interviewed Thai farmers use antibiotics in aquaculture. The farmers in the study used on average 13 chemicals and biological products in shrimp pond management. In comparison with shrimp farms from Thailand, farms from Mexico used an average 41.7 chemical and biological products in each farm (Lyle-Fritch et al., 2006). Sapkota et al. (2008) conducted a review of the 26 antibiotics as listed from the FAO and found that oxytetracycline, chloramphenicol, and oxolinic acid usage tops the list between 1990 and 2007.

Antibiotic residues such as tetracyclines, fluoroquinolones and sulfonamides were found in shrimp tissues (Dang et al., 2010; Love et al., 2011). In fact most veterinary drug violations were detected in species that are commonly farmed. Asian seafood products such as shrimps, catfish, crab, tilapia, eel, and Chilean salmon were most frequently found to violate veterinary drug residue standards (Love et al., 2011). The increased usage of antibiotics may lead to increased levels of antibiotic resistance bacteria. For example, in a study conducted in shrimp ponds in the Philippines, Tendencia and Peña (2001) found that the prevalence of multiple antibiotic resistance among *Vibrio* spp. was highest in shrimp ponds treated with oxolinic acid compared to untreated ponds. *Aeromonas* spp. was isolated from retail fish samples in Malaysia and all isolates demonstrated resistance to three or more antibiotics (Radu et al., 2003). The presence of antibiotic residues in aquaculture environments and products could result in adverse ecological and public health effects. Low-level exposures to antibiotic residues present in the environment and food are not likely to cause acute toxic effects among the general public; however, chronic effects are still largely unstudied (Jones et al., 2004).

7.2.1.3 Wastewater and Run-Offs

The contamination of aquaculture ponds with wastewater and animal waste may be increasing due to surface water pollution. Besides unintentional contamination, some Asian countries such as Vietnam uses excreta in fish farming; sewage in aquaculture ponds remains the available method of wastewater treatment and disposal in some northern parts of Vietnam (WHO, 2006). Integrated fish farming, which combines livestock production with fish farming, is practiced throughout Southeast Asia as well. Animal manure is shed directly into fish ponds and directly consumed by the fish. The livestock, mainly chickens and pigs, are often fed feed containing growth promoters and antibiotics. This causes an increased resistance toward antibiotics in *Acinetobacter* spp. and *Enterococcus* spp. isolated from the fish (Petersen et al., 2002). Research studies by Lyle-Fritch et al. (2006) revealed that 39% of 23 Mexican shrimp farms' water supply was contaminated with agricultural run-offs and effluents from other shrimp farms. Agrochemicals such as pesticides, antifungals, disinfectants, fertilizers, and other water treatment compounds are also used in aquaculture. Run-off from agricultural sites can serve as potential agrochemical contamination into aquaculture ponds (Gräslund and Bengtsson, 2001).

7.2.2 Intervention Strategies

7.2.2.1 HACCP at the Farm Level

At the current stage of research, a true HACCP plan may be possible for chemical hazards (MacDonald, 2005), but risk reduction points or farm specific pathogen reduction programs may be more suitable for microbiological hazards in the salmon farm. The HACCP system may be relevant to the salmon farm because some controls are more effective at the farm level, that is, veterinary residues, organochlorine (e.g., dioxins, PCBs, and PBDEs), and pesticides (e.g., endosulfan from plant-based feed) can generally be controlled most effectively on the farm (Figure 7.2).

7.2.2.2 Site Selection

Hazards may exist with the location of the fish site. Fish farms can be subjected to pesticide and chemical run-off from adjacent agricultural land or industrial sources and leads to unacceptable levels of chemical contaminants in the product (Reilly and Käferstein, 1997). Some of the main sources of heavy metal contamination found in coastal waters where fish and shellfish are farmed include industrial and municipal waste discharge, antifouling

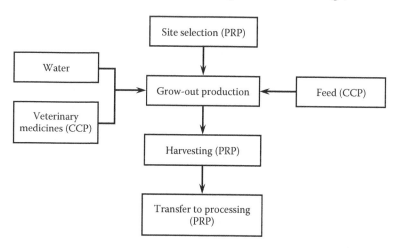

FIGURE 7.2
HACCP-based plan for salmon production at marine sites. (CCP: critical control point; PRP: prerequisite programs).

paints, and various organic pesticides, herbicides, and hydrocarbons. Exposure to these sources can be minimized by licensing farming areas away from sources of contamination (Fairgrieve and Rust, 2003). The discharge of effluents into the environment may affect fish stocks. For example, River Don in Aberdeenshire, Scotland, has supported a modest salmon fishery, but the presence of dead or dying yellow, orange, and red pigmented wild broodfish threatened the fishery industry. Affected fish were predominantly female and pigmented from yellow through to red mainly along the ventral surface, fin bases, and in the cartilage surrounding the eyes. The pigmentation resulted from hemosiderin (yellow to brown iron-rich pigments) deposition following exposure to pulp mill effluents (Bruno and Poppe, 1996; Everall et al., 1992). The choice of locality should be based on the environmental conditions which give the best possible conditions for fish growth, while avoiding areas near to industrial, agricultural, and urban areas.

7.2.2.3 Feed/Medicated Feed

Diet is the main source of exposure to a wide range of contaminants in fish. Fish oil is the main source of persistent organic pollutants while fish meal is the main source of elements such as mercury, cadmium, and arsenic (Berntssen et al., 2010). In farmed fish, the level of contaminants in feed materials can be monitored and controlled (EFSA, 2005; Jacobs et al., 2002b); hence, feed is suggested as a potential Critical Control Point (CCP) in salmon production. All feed suppliers should supply farms with a written confirmation that the feed conformed to safety and quality requirements and had been tested for contaminants such as polychlorinated biphenyls (PCBs), dioxins, polybrominated diphenyl ethers (PBDEs), organochlorine pesticides, and methyl mercury.

All veterinary medicines for salmon are prescription-only drugs, and the farm veterinarian has the responsibility for prescribing the most effective treatment regime. Upon receipt of a prescription for a medicated feeding stuff, the feed manufacturer may coat antibiotics, and other medicines licensed for in-feed use, and pellets for in-feed administration. The manufacture of medicated feed can be labor intensive, which typically takes place on a dedicated feed line. It is essential that cross-contamination between different medicines and between nonmedicated and medicated feed cannot occur. Between each batch of a different medication every part of the line must be cleaned. Each batch of feed must be sampled to ensure that the medication is present as prescribed and distributed effectively in the feed.

Farm operators can avoid the accidental feeding of medicated feed to the wrong fish by using distinct colored labeling and packaging, which clearly identifies the medicated product. It is important to ensure that all fish to be treated receive the medicine, and it is normal to add the medicine to between 50–75% of the daily ration. It is illegal to sell fish containing residues of a medicine that exceed a certain legal maximum level. In order to assist this process, each licensed medicine is issued with a withdrawal period. The withdrawal period is typically described in degree days (e.g., 50° days are 5 days at 50°F/10°C or 10 days at 4°F/5°C). Withdrawal period is the time that must pass after the last dose of drug administration and the time that the fish is slaughtered for human consumption to ensure that harmful residues will not be present in edible fish parts. It is calculated to ensure that any medicinal residue in the edible tissues will be below the legal limit designated for that medicine at the time the fish is harvested for consumption (Sinnott, 2002).

7.2.2.4 Water

The growing-out phase of salmon is undertaken in seawater. According to Maage et al. (2008), the quality of seawater supply does not have a great impact on the food safety issues as compared to feed.

7.2.2.5 Harvesting and Transfer to the Processing Plant

Harvest bins are in good condition with sealed bin liners and secure lids and bindings. Harvest bins are thoroughly cleaned and disinfected after harvesting. Onsite harvesting equipment should be site-specific and if moved between sites should be disinfected. When harvesting, it is necessary to reduce the core temperature of fish below 5°C by the time they arrive at the packing station. This is important to inhibit bacterial growth in the harvested fish.

7.2.2.6 Biosecurity Measures on Farm

Not knowing the extent to which biosecurity measures need to be employed to prevent the transmission of pathogens and infectious diseases is an important problem (Amass et al., 2000), because, until that information is known, producers will run one of two risks:

- Expend funds on unnecessary biosecurity measures
- Continue insufficient biosecurity measures that may result in outbreaks

Figure 7.3 shows the biosecurity procedures carried out before workers or visitors enter the salmon farms. Boats, boots, and hands are disinfected at different stages to prevent or reduce potential cross-contamination of infectious agents to the farms.

Specific Pathogen Free (SPF) shrimp can be used in farms to mitigate production loss from disease. SPF shrimp are free of one or more specific pathogens which meet the following criteria: (i) the pathogen can be reliably diagnosed; (ii) the pathogen can be physically excluded from a facility,

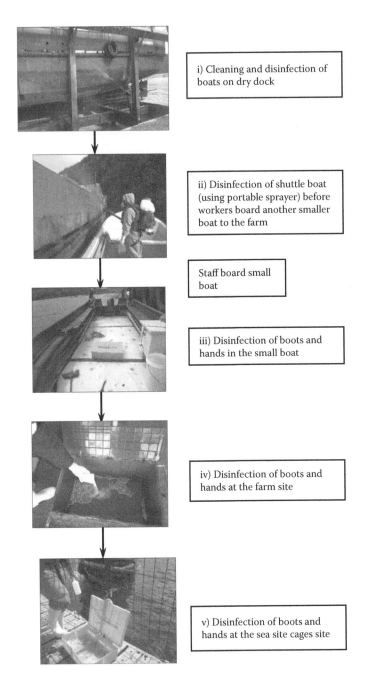

FIGURE 7.3
Disinfection procedures before entering salmon sea sites.

FIGURE 7.4
Bird nettings and crab barriers set up at a shrimp farm.

and (3) the pathogen poses a significant threat to the industry. There are SPF populations of shrimp which are free of white spot syndrome virus and yellow head virus (Lightner et al., 2009). Although pathogens such as *Vibrio* spp. pose a significant threat to the industry and can be reliably diagnosed, they do not meet the third specification of SPF since they cannot be physically excluded from a facility due to its ubiquity in a shrimp's normal gut flora (Moss et al., 2012). On farm biosecurity practices which include influent water quality, treatment of water effluent, fallowing of ponds before restocking, usage of plastic-lined ponds, crab fencing and bird nettings (Figure 7.4) to prevent infectious diseases should be implemented. Quarantine, control of traffic personnel, vehicles and equipment, vaccination and/or medication, depopulations and eradication are other crucial biosecurity protocols to prevent and control disease (Pruder, 2004). Recently, a shrimp farm in Malaysia has successfully redesigned their farm to combine biosecurity and biofloc technology with minimum water exchange (Taw et al., 2011). Bioflocs are single-celled proteins produced in biological reactors and can be used as an alternative for soy and fishmeal proteins in shrimp diets. The biological process converts dissolved nutrients (nitrogen and phosphorus) into proteins contained within bioflocs (Kuhn et al., 2011).

7.2.2.7 Risk Assessment

Various risk assessments such as health and safety, control of substances hazardous to health, workers' hygiene, biosecurity measures, veterinary medicine, fish health, feed safety, and site assessment can be carried out regularly. Table 7.3 shows some examples of risk assessments conducted at the farm level. The risk assessments are carried out to assess the performance of various farm sites ranging from food safety and biosecurity measures to emergency responses. In addition to public and fish health, fish welfare and productivity losses were taken into consideration as well.

The probability of occurrences were ranked on a scale of 1 to 5, here, 1 = very unlikely to occur and 5 = very likely to occur. Severity scoring is also ranked on a scale of 1 to 5, where 1 = "not very severe" and 5 = "very severe." When risk was calculated (i.e., probability scoring × severity scoring), the maximum risk weight = 25. When the risk assessments were completed, a total risk weight is calculated. Table 7.3 shows nine examples of risk assessment exercises, hence, the total risk weight for a particular site is 25 × 9 = 225. The risk assessment format provided a risk weighting, scaled between 0 and 225, where 0 represents no risk (which is unlikely) and 225 represents the opposite extreme.

The risk weighting can further be categorized into:

0–45: Very low risk farm.

46–90: Low risk farm; some areas may require evaluation to determine areas for improvements.

91–135: Medium risk farm; some areas may require evaluation to determine areas for improvements. Reassess after corrective actions have been implemented.

136–180: High risk farm; areas needing urgent evaluation and corrective actions. Immediate actions to be taken and to reassess after preventive and corrective measures have been implemented.

181–225: Very high risk farm; areas needing urgent and immediate corrective actions. Immediate actions need to be taken and to reassess after preventive and corrective measures have been implemented.

The risk assessment exercises provide guidance on the allocation of specific salmon farm sites into categories based on their likelihood of contributing hazards (i.e., food safety and infectious disease threat) and the potential magnitude of that contribution. It served as a valuable communication tool to categorize different farm sites so that the management can give greater attention to farms with the greatest potential risks to fish and public health. Farmers were encouraged to identify potential hazards on the farm and to develop an appropriate action plan for improvement; the farm personnel and management can determine the most vulnerable points and focus resources on preventing and reducing risks on the farm.

TABLE 7.3

Risk Assessments Conducted at Sea Sites

Parameters	Probability[a]	Justification	Severity[b]	Justification	Risk Weight
Food safety					
• Chemical contaminants from feed	1	Certificate of analysis from approved supplier	3	Reject and recall salmon; productivity loss	3
Diseases					
• Infectious Pancreatic Necrosis	1	Under control	3	Can result in losses; transmitted through water and eggs	3
• Infectious Salmon Anemia	3	Potential for infectious diseases (no vaccines—dependent on biosecurity measures)	5	ISA. Production loss; spread of disease through water and live fish	15
• Piscirickettsia salmonis	3	Main problem in Chile	5	May trigger ISA	15
Water quality					
• Fuel spillage	1	Risk assessment of boat; sensory analyses of salmon (no fuel smell/taste)	4	Rejected and recalled; production losses; loss of customer confidence	4
Predators (sea lions)	1	Anti-sea lions net; no previous attacks	3	Production losses from attack and escape fish; additional stress	3
Emergency response					
• Fire hazard	1	Trained staff; fire extinguishers in place	3	Staff injury; production loss	3
• Chemical spillage	1	Cemented area to maintain spillage	3	Staff injury	3
• Earthquake	1	Review of trend and forecasting; study of seismic waves; emergency response training	4	Production losses; farm closed down	4

Note: Maximum scoring = 225 (highest probability ranking = 5; highest severity scoring = 5; hence, 9 risk assessments = $25 \times 9 = 225$).

[a] Probability scoring on a scale of 1–5, where 1 = not very likely to occur; 5 = very likely to occur.

[b] Severity scoring on a scale of 1–5, where 1 = not very serious consequences; 5 = very serious consequences.

7.2.2.8 Good Manufacturing Practices (GMP) and HACCP at the Processing Plant

Fish should be mixed with an adequate amount of ice to keep it chilled during storage and distribution. About 2.5 kg of ice are required to cool 10 kg of salmon at about 59°F (15°C) to ice temperature. It is not easy to estimate the extra amount required to keep the fish chilled during subsequent distribution and storage; the degree of insulation of the transport boxes, the insulation of refrigeration of the carrying vehicle or shipping container, and the length of the distribution chain are important factors. Experience is a better guide than trying to estimate these factors, but generally salmon would be iced in a ratio of 3:1 to 2:1 (fish to ice) for a journey of a few days in insulated or refrigerated shipping containers or vehicles (Howgate, 2002). Ice should be manufactured from potable water or obtained from certified supplier. Ice should make good contact with fish, avoidance of physical damage, and ease of application. Flake ice meets this requirement; its small, flat plates make good contact with the surface of the fish, rapidly cooling, and do not damage the skin of the salmon. Another form of ice that has some advantages is slush ice. This is prepared from water to which some salt has been added and is prepared as a slurry of ice in dilute brine. This material can be pumped, makes excellent contact with the fish, and does not cause any physical damage, apart perhaps of some dulling of the appearance of the skin. When first prepared, slush ice has a temperature of a little below 32°F (0°C) because the salt lowers the freezing temperature of the mix. This lower temperature slows the rate of spoilage slightly compared with storage in normal ice, but this advantage wears off during storage as the ice melts and the brine drains away (Howgate, 2002). Fish should be stored under refrigeration temperature 30.2°F to 41°F (−1°C to 5°C) to avoid potential microbiological growth (DAFF, 2005).

L. monocytogenes is an environmental pathogen and may be present in cooling units, air handling equipment, standing/dripping water, or condensation. The moist environment is conducive to *L. monocytogenes* growth. Immediately address and correct problems of dripping, condensation, and standing water. Cooling units and air-handling units should be cleaned at specific intervals. It can be stated that the control options for *L. monocytogenes* during processing is primarily a matter of having a proper cleaning and sanitation program. A specific hygiene and sanitation program needs to be developed in order to keep the contamination with *L. monocytogenes* at a low level (Huss et al., 2000; Rørvik, 2000). Cleaning processing equipment, food contact surfaces, separation of staff functions, personal hygiene, and restrictions on entry of visitors are important to reduce the prevalence of *L. monocytogenes* (Garland, 1995). Autio et al. (1999) conducted an eradication program for decreasing the level of *L. monocytogenes* contamination. Skinning, slicing and brining machines (in a cold-smoked processing plant) were disassembled and thoroughly cleaned and disinfected and the pieces were either placed in hot water or heated in an oven (176°F [80°C]). The production line, floors, and walls were treated thoroughly with hot steam. During the heat treatment program, no *L. monocytogenes* were isolated in 188 control samples taken during the 5-month period when heat treatment was applied. A separate hygiene and sanitation team could be hired to work when processing finishes (e.g., from 5 pm to 2 am). This is to ensure that the processing staff are not too burdened with both processing and washing duties and the hygiene team can specifically concentrate on cleaning and disinfection of the processing plant.

Adequate cleaning and disinfection must be carried out as *L. monocytogenes* may persist in the processing environment. Vogel et al. (2001) found that a similar *L. monocytogenes* strain was found over a 4-year period indicating that the *L. monocytogenes* has established and persisted as an in-house flora and was not eliminated by routine hygienic procedures. Similarly in another study, *L. monocytogenes* was found in 34.6% of samples taken from floors during shrimp processing and in 15.4% of samples obtained after cleaning (Gudmundsdóttir et al., 2006). Bacterial cells can be displaced by liquids or aerosols caused by mechanical action during processing and cleaning operations (e.g., scrubbing, washing with low pressure hoses or high pressure jets) and may be protected in harborage sites. Water and organic matter may become stagnate in harborage sites (i.e., niches or hard-to-reach places) allowing for microbial growth (Carpentier and Cerf, 2011). Some examples of hard- or difficult-to-clean sites are hollow parts, unpolished welds, new materials (e.g., gaskets) or worn materials (e.g., cracks on floors or conveyor belts' materials), and corroded spots (Carpentier, 2005). This indicates the importance of cleaning and disinfection of processing plant environments to reduce contamination of final seafood products.

Similar to fish, the processing of shrimp products must operate under HACCP guidelines. The microbial hazard identified are high bacterial load associated with raw material, cross-contamination from workers during handling, microbial growth due to improper freezing, contamination from recirculated processing water, or glaze water and cross-contamination of finished product during packing. Microbial load associated with raw material can be neutralized by going through a preliminary wash in 50 ppm chlorine water and then through a pneumatic filth washing machine in order to remove filth attached to the shrimp. Cross-contamination during peeling and other handling steps can be kept under check by wearing sterile gloves and periodical hand dip in chlorinated water maintained at 50 ppm. A nose-piece/mask should be worn correctly. The peeling knife and utensils are alls rinsed periodically in chlorinated water and peeled-off waste material is removed regularly to a large polyethylene bag kept outside the plant through a chute. The freezer temperature are constantly monitored and maintained at −°F to −40°F (−35°C to −40°C). Recirculated glaze water should be controlled through proper chlorination and passed through UV filters (Hatha et al., 2003).

References

Akinbowale, O. L., H. Peng, and M. D. Barton. 2006. Antimicrobial resistance in bacteria isolated from aquaculture sources in Australia. *J. Appl. Microbiol.* 100: 1103–1113.

Alam, M. G. M., A. Tanaka, G. Allinson, L. J. B. Laurenson, F. Stagnittim, and E. T. Snow. 2002. A comparison of trace element concentrations in cultured and wild carp (*Cyprinus carpio*) of Lake Kasumigaura, Japan. *Ecotoxicol. Environ. Saf.* 53: 348–354.

Amagliani, G.,G. Brandi, and G. F. Schiavano. 2012. Incidence and role of *Salmonella* in seafood safety. *Food Res. Int.* 45: 780–788.

Amass, S.F., B.D. Vyverberg, D. Ragland et al. 2000. Evaluating the efficacy of boot baths in biosecurity protocols. *Swine Health Prod.* 8: 169–173.

Angot, V. and P. Brasseur. 1993. European farmed Atlantic salmon (*Salmo salar* L.) are safe from anisakid larvae. *Aquaculture* 118: 339–344.

Autio, T., S. Hielm, M. Miettinen et al. 1999. Sources of *Listeria monocytogenes* contamination in a cold-smoked rainbow trout processing plant detected by pulsed-field gel electrophoresis typing. *Appl. Environ. Microbiol.* 65: 150–155.

Autio, T., T. Säteri, M. Fredriksson-Ahomaa, M. Rahkio, J. Lundén, and H. Korkeala. 2000. *Listeria monocytogenes* contamination pattern in pig slaughterhouses. *J. Food Prot.* 63: 1438–1442.

Beaufort, A., S. Rudelle, N. Ganou-Besse et al. 2007. Prevalence and growth of *Listeria monocytogenes* in naturally contaminated cold-smoked salmon. *Lett. Appl. Microbiol.* 44: 406–411.

Bell, J. G., F. McGhee, J. R. Dick and D. R. Tocher. 2005. Dioxin and dioxin-like polychlorinated biphenyls (PCBs) in Scottish farmed salmon (*Salmo salar*): Effects of replacement of dietary marine fish oil with vegetable oils. *Aquaculture* 243: 305–314.

Berntssen, M. H. G., C. N. Glover, D. H. F. Robb, J.-V. Jakobsen, and D. Petri. 2008. Accumulation and elimination kinetics of dietary endosulfan in Atlantic salmon (*Salmo salar*). *Aquat. Toxicol.* 86: 104–111.

Berntssen, M. H. G. and A.-K.Lundebye. 2008. Environmental contaminants in farmed fish and potential consequences for seafood safety. In *Improving Farmed Fish Quality and Safety*, ed. L. Øyvind, 39–70. Cambridge: Woodhead Publishing Ltd.

Berntssen, M. H. G., K. Julshamn, and A.-K. Lundebye. 2010. Chemical contaminants in aquafeeds and Atlantic salmon (*Salmo salar*) following the use of traditional – versus alternative feed ingredients. *Chemosphere* 78: 637–646.

Bhaskar, N. and N. M. Sachindra. 2006. Bacteria of public health significance associated with cultured tropical shrimp and related safety issues: A review. *J. Food Sci. Techol. Mys.* 43: 228–238.

Bhaskar, N., T. M. R. Setty, G. V. S. Reddy et al. 1995. Incidence of *Salmonella* in cultured shrimp *Penaeus monodon. Aquaculture* 138: 257–266.

Blanco, S. L., C. Sobrado, C. Quintela, S. Cabaleiro, J. C. González, and J. M. Vieites. 2007. Dietary uptake of dioxins (PCDD/PCDFs) and dioxin-like PCBs in Spanish aquacultured turbot (*Psetta maxima*). *Food Addit. Contam.* 24: 421–428.

Bocio, A., J. L. Domingo, G. Falcó, and J. M. Llobet. 2007. Concentrations of PCDD/ PCDFs and PCBs in fish and seafood from the Catalan (Spain) market: Estimated human intake. *Environ. Int.* 33: 170–175.

Bodensteiner, L. R., R. J. Sheehan, P. S. Wills, A. M. Brandenburg, and W. M. Lewis. 2000. Flowing water: An effective treatment for Ichthyophthiriasis. *J. Aquat. Anim. Health* 12: 209–219.

Bondad-Reantaso, M. G., R. P. Subasinghe et al. 2005. Disease and health management in Asian aquaculture. *Vet. Parasitol.* 132: 249–272.

Bruno, D. and T. T. Poppe. 1996. A colour atlas of salmonid diseases, 1–194. London: Academic Press.

Cabanillas-Beltrán, H., E. Llausás-Magaña, R. Romero et al. 2006. Outbreak of gastroenteritis caused by the pandemic *Vibrio parahaemolyticus* O3:K6 in Mexico. *FEMS Microbiol. Lett.* 265: 76–80.

Carpentier, B. 2005. Improving the design of floors. In *Handbook of Hygiene Control in the Food Industry*, ed. H. L. M. Lelieveld, M. A. Mostert, and J. Holah, 168–184. Cambridge: Woodhead Publishing Ltd.

Carpentier, B. and O. Cerf. 2011. Review – Persistence of *Listeria monocytogenes* in food industry equipment and premises. *Int. J. Food Microbiol.* 145: 1–8.

Chapman, P. M., H. E. Allen, K. Godtfredsen, and M. N. Z'Graggen. 1996. Evaluation of bioaccumulation factors in regulating metals. *Environ. Sci. Technol.* 30: 448–452.

Chou, C. L. 2007. A time series of mercury accumulation and improvement of dietary feed in net caged Atlantic salmon (*Salmo salar*). *Mar. Pollut. Bull.* 54: 720–725.

DAFF. 2005. Hazard analysis critical control point (HACCP): A guideline to compliance with the export control (fish & fish products) orders. http://www.daff.gov.au/__data/assets/pdf_file/0019/126181/haccp_ffp.pdf (accessed October 15, 2010).

Dang, P. K., G. Degand, S. Danyi et al. 2010. Validation of a two-plate microbiological method for screening antibiotic residues in shrimp tissue. *Anal. Chim. Acta* 672: 30–39.

Darnerud, P. O., S. Atuma, M. Aune et al. 2006. Dietary intake estimations of organohalogen contaminants (dioxins, PCB, PBDE and chlorinated pesticides, e.g., DDT) based on Swedish market based data. *Food Chem. Toxicol.* 44: 1597–1606.

DePaola, A., J. L. Nordstrom, J. C. Bowers, J. G. Wells, and D. W. Cook. 2003. Seasonal abundance of total and pathogenic *Vibrio parahaemolyticus* in Alabama oysters. *Appl. Environ. Microbiol.* 69: 1521–1526.

Dewailly, É., P. Ayotte, M. Lucas, and C. Blanchet. 2007. Risks and benefits from consuming salmon and trout: A Canadian perspective. *Food Chem. Toxicol.* 45: 1343–1348.

Directive 2002/32/EC of the European Parliament and of the Council of 7 May 2002 on undesirable substances in animal feed. *Off. J. Eur. Comm.* L 140: 10–21. http://eur-lex.europa.eu/pri/en/oj/dat/2002/l_140/l_14020020530en00100021.pdf (accessed April 29, 2010).

Dorval, J.,V. S. Leblond, and A. Hontela. 2003. Oxidative stress and loss of cortisol secretion in adrenocortical cells of rainbow trout (*Oncorhynchus mykiss*) exposed in vitro to endosulfan, an organochlorine pesticide. *Aquat. Toxicol.* 63: 229–241.

Duncan, P. F. 2003. Aquaculture of commercially important molluscs and crustaceans. In, *Encyclopedia of Food Sciences and Nutrition*, ed. B. Caballero, L. Trugo, and P. M. Finglas, 5245–5251. Amsterdam: Elsevier Science Ltd.

Eagle, J., R. Naylor and W. Smith. 2004. Why farm salmon outcompete fishery salmon. *Mar. Policy* 28: 259–270.

Easton, M. D. L., D. Luszniak, and E. Von der Geest. 2002. Preliminary examination of contaminant loadings in farmed salmon and commercial salmon feed. *Chemosphere* 46: 1053–1074.

EC. 2006. Commission Regulation (EC) No 199/2006 of 3 February 2006 amending Regulation (EC) No 466/2001 setting maximum levels for certain contaminants in foodstuffs as regards dioxins and dioxin like PCBs. *Off. J. Eur. Union* L32: 34–38. http://eur-lex.europa.eu/LexUriServ/LexUriServ.do?uri=OJ:L:2006:03 2:0034:0038:EN:PDF (accessed March 29, 2012).

EFSA. 2005. European Food Safety Authority. Opinion of the Scientific Panel on Contaminants in the Food Chain on a Request from the European Parliament Related to the Safety Assessment of Wild and Farmed Fish. *EFSA Journal* 236: 1–118. http://www.efsa.europa.eu/en/scdocs/doc/236.pdf (accessed April 24, 2010).

Espe, M., A. Lemme, A. Petri, and A. El-Mowafi.2007. Assessment of lysine requirement for maximal protein accretion in Atlantic salmon using plant protein diets. *Aquaculture* 263: 168–178.

EU. 2000. European Union risk assessment report. Diphenyl ether, pentabromo derivative (Pentabromodiphenyl ether). Report No.: CAS No.: 32534-81-9, EINECS No.: 251-084-2.

EU. 2002. European Union risk assessment report. Bis(pentabromophenyl) ether. Report No.: CAS No.: 1163-19-5, EINECS No.: 214-604-9.

EU. 2003. European Union risk assessment report. Diphenyl ether, octabromo derviative. Report No.: CAS No.: 32536-52-0, EINECS No.: 251-087-9.

Everall. N. C., C. G. Mitchell, and J. N. Robson. 1992. Effluent causes of the "pigmented salmon syndrome" in wild adult Atlantic salmon (*Salmo salar*) from the River Don in Aberdeenshire. *Dis. Aquat. Organ.* 12: 199–205.

Fairgrieve, W. T. and M. B. Rust. 2003. Interactions of Atlantic salmon in the Pacific northwest V. Human health and safety. *Fish. Res.* 62: 329–338.

FAO. 2001. Hazard identification, exposure assessment and hazard characterization of *Campylobacter* spp. in broiler chickens and *Vibrio* spp. in seafood. Joint FAO/WHO Expert Consultations on risk assessment of microbiological hazards in foods, Geneva, Switzerland, 23–27 July 2001. http://www.fao.org/docrep/008/ae521e/ae521e00.htm (accessed February 13, 2010).

FAO. 2006. State of world aquaculture 2006. FAO Fisheries Technical Paper No. 55. FAO, Rome, 1–134. ftp://ftp.fao.org/docrep/fao/009/a0874e/a0874e00.pdf (accessed April 13, 2010).

FAO. 2008. Glossary of aquaculture. http://www.fao.org/fi/glossary/aquaculture/ (accessed March 26, 2012).

FAO. 2009a. Aquaculture. In *CWP Handbook of Fishery Statistical Standards*.http://www.fao.org/fishery/cwp/handbook/J/en (accessed September 28, 2009).

FAO. 2009b. The state of world fisheries and aquaculture 2008. Food and Agriculture Organization of the United Nations, Rome. ftp://ftp.fao.org/docrep/fao/011/i0250e/i0250e.pdf (accessed February 9, 2010).

FAO. 2009c. World aquaculture production by species group. Yearbook of fishery statistics summary table.ftp://ftp.fao.org/FI/STAT/summary/b-1.pdf (accessed March 26, 2012).

FAO. 2010. World aquaculture production of fish, crustaceans, mollusks, etc., by principal species in 2007. Statistics. FAO: Fisheries and Aquaculture Department. http://www.fao.org/fishery/statistics/en (accessed May 1, 2010).

FAO. 2012. Trends in the fisheries sector. FAO Statistical Yearbook 2012. http://www. fao.org/docrep/015/i2490e/i2490e00.htm (accessed March 26, 2012).

FISHSTAT. 1998. Aquaculture production by species group. Yearbooks of fishery statistics. ftp://ftp.fao.org/fi/stat/summary/summ_98/aq_b0.pdf (accessed April 13, 2010).

FISHSTAT. 2007. World aquaculture production by species group. Yearbooks of fishery statistics. ftp://ftp.fao.org/fi/stat/summary/b-1.pdf (accessed April 13, 2010).

Foran, J. A., D. O. Carpenter, M. C. Hamilton, B. A. Knuth, and S. J. Schwager. 2005. Risk-based consumption advice for farmed Atlantic and wild pacific salmon contaminated with dioxins and dioxin-like compounds. *Environ. Health Perspect.* 113: 552–556.

Garland, C. D. 1995. Microbial quality of aquaculture products with special reference to *Listeria monocytogenes* in Atlantic salmon. *Food Aust.* 47: 550–563.

Glover, C. N., D. Petri, K.-E.Tollefsen, N. Jørum, R. D. Handy and M. H. G. Berntssen. 2007. Assessing the sensitivity of Atlantic salmon (*Salmo salar*) to dietary endosulphan exposure using tissue biochemistry and histology. *Aquat. Toxicol.* 84: 346–355.

Gopal, S., S. K. Otta, S. Kumar, I. Karunasagar, M. Nishibuchi, and I. Karunasagar. 2005. The occurrence of *Vibrio* species in tropical shrimp culture environments; implications for food safety. *Int. J. Food Microbiol.* 102: 151–159.

Gräslund, S. and B.-E. Bengtsson. 2001. Chemicals and biological products used in south-east Asian shrimp farming, and their potential impact on the environment – a review. *Sci. Total Environ.* 280: 93–131.

Gräslund, S., K. Holmström, and A. Wahlström. 2003. A field survey of chemicals and biological products used in shrimp farming. *Mar. Pollut. Bull.* 46: 81–90.

Gudmundsdóttir, S., B. Gudbjörnsdóttir, H. Einarsson, K. G. Kristinsson, and M. Kristjánsson. 2006. Contamination of cooked peeled shrimp (*Pandalus borealis*) by *Listeria monocytogenes* during processing at two processing plants. *J. Food Prot.* 69: 1304–1311.

Hatha, A. A. M., T. L. Maqbool, and S. S. Kumar. 2003. Microbial quality of shrimp products of export trade produced from aquaculture shrimp. *Int. J. Food Microbiol.*82: 213–221.

Hayward, D., J. Wong, and A. J. Krynitsky. 2007. Polybrominated diphenyl ethers and polychlorinated biphenyls in commercially wild caught and farm-raised fish fillets in the United States. *Environ. Res.* 103: 46–54.

Hites, R. A., J. A. Foran, D. O. Carpenter, M. C. Hamilton, B. A. Knuth, and S. J. Schwager. 2004a. Global assessment of organic contaminants in farmed salmon. *Science* 303: 226–229.

Hites, R. A., J. A. Foran, S. J. Schwager, B. A. Knuth, M. C. Hamilton, and D. O. Carpenter. 2004b. Global assessment of polybrominated diphenyl ethers in farmed and wild salmon. *Environ. Sci. Technol.* 38: 4945–4949.

Holmström, K., S. Gräslund, A. Wahlström, S. Poungshompoo, B.-E. Bengtsson and N. Kautsky. 2003. Antibiotic use in shrimp farming and implications for environmental impacts and human health. *Int. J. Food Sci. Technol.* 38: 255–266.

Horst, K., I. Lehmann, and K. Oetjen. 1998. Levels of chlorine compounds in fish muscle, -meal, -oil and -feed. *Chemosphere* 36: 2819–2832.

Howgate, P. 1998. Review of the public health safety products from aquaculture. *Int. J. Food Sci. Technol.* 33: 99–125.

Howgate, P. 2002. Postharvest handling and processing. In *Handbook of Salmon Farming*, ed. S. M. Stead and L. Laird, 187–202. Chichester: Springer Praxis.

Huss, H. H., P. K. B. Embarek, and V. F. Jeppesen. 1995. Control of biological hazards in cold smoked salmon production. *Food Control* 6: 335–340.

Huss, H. H., A. Reilly, and P. K. B. Embarek. 2000. Prevention and control of hazards in seafood. *Food Control* 11: 149–156.

Jacobs, M., A. Covaci, and P. Schepens. 2002a. Investigation of selected persistent organic pollutants in farmed Atlantic salmon (*Salmo salar*), salmon aquaculture feed, and fish oil components of the feed. *Environ. Sci. Technol.* 36: 2797–2805.

Jacobs, M., J. Ferrario, and C. Byrne. 2002b. Investigation of polychlorinated dibenzo-*p*-dioxins, dibenzo-*p*-furans and selected coplanar biphenyls in Scottish farmed Atlantic salmon (*Salmo salar*). *Chemosphere* 47: 183–191.

Jang, C.-S., C.-W. Liu, K.-H. Lin, F.-M. Huang, and S.-W. Wang. 2006. Spatial analysis of potential carcinogenic risks associated with ingesting arsenic in aquacultural tilapia (*Oreochromic mossambicus*) in Blackfoot disease hyperendemic areas. *Environ. Sci. Technol.* 40: 1701–1713.

Johansson, T., L. Rantala, L. Palmu, and T. Honkanen-Buzalski. 1999. Occurrence and typing of *Listeria monocytogenes* strains in retail vacuum-packed fish products and in a production unit. *Int. J. Food Microbiol.* 47: 111–119.

Jones, O., N. Voulvoulis, and J. Lester. 2004. Potential ecological and human health risks associated with the presence of pharmaceutically active compounds in the aquatic environment. *Crit. Rev. Toxicol.* 34: 335–350.

Jørgensen, L. V. and H. H. Huss. 1998. Prevalence and growth of *Listeria monocytogenes* in naturally contaminated seafood. *Int. J. Food Microbiol.* 42: 127–131.

Kelly, B. C., M. P. Fernandez, M. G. Ikonomou, and W. Knapp. 2008. Persistent organic pollutants in aquafeed and Pacific salmon smolts from fish hatcheries in British Columbia, Canada. *Aquaculture* 285: 224–233.

Kiviranta, H., M.-L. Ovaskainen, and T. Vartiainen. 2004. Market basket study on dietary intake of PCDD/Fs, PCBs, and PCDEs in Finland. *Environ. Int.* 30: 923–932.

Knowles, T. G., D. Farrington, and S. C. Kestin. 2003. Mercury in UK imported fish and shellfish and UK-farmed fish and their products. *Food Addit. Contam.* 20: 813–818.

Kodama, H., Y. Nakanishi, F. Yamamoto et al. 1987. *Salmonella arizonae* isolated from a pirarucu, *Arapaima gigas* Cuvier, with septicaemia. *J. Fish Dis.* 10: 509–512.

Koonse, B. 2006. Seafood safety: Down on the farm. In *Food Safety Magazine*, ed. C. Ainsworth. http://www.foodsafetymagazine.com/article.asp?id=565&sub=sub1#Koonse (accessed September 25, 2009).

Kuhn. D. D., A. L. Lawrence, and G. D. Boardman. 2011. What makes bioflocs great for shrimp? *Global Aquacul. Adv.* (March/April): 76–77.

Lappi, V. R., A. Ho, K. Gall, and M. Wiedmann. 2004. Prevalence and growth of *Listeria* on naturally contaminated smoked salmon over 28 days of storage at 4°C. *J. Food Prot.* 67: 1022–1026.

Legrand, M., P. Arp, C. Ritchie, and H. M. Chan. 2005. Mercury exposure in two coastal communities of the Bay of Fundy, Canada. *Environ. Res.* 98: 14–21.

Lightner, D. V., R. M. Redman, S. M. Arce, and S. M. Moss. 2009. Specific pathogen free shrimp stocks in shrimp farming facilities as a novel method for disease control in crustaceans. In *Shellfish Safety and Quality*, ed. S. E. Shumway and G. E. Rodrick, 384–424. Cambridge: Woodhead Publishing Ltd.

Lillehaug, A., B. T. Lunestad, and K. Grave. 2003. Epidemiology of bacterial diseases in Norwegian aquaculture—a description based on antibiotic prescription data for the ten-year period 1991 to 2000. *Dis. Aquat. Organ.* 53: 115–125.

Little, D. C., G. K. Milwain, and C. Price. 2008. Pesticide contamination in farmed fish: Assessing risks and reducing contamination. In *Improving Farmed Fish Quality and Safety*, ed. L. Øyvind, 70–96. Cambridge: Woodhead Publishing Limited.

Liu, Y. and U. R. Sumaila. 2008. Can farmed salmon production keep growing? *Mar. Policy* 32: 497–501.

Love, D. C., S. Rodman, R. A. Neff, and K. E. Nachman. 2011. Veterinary drug residues in seafood inspected by the European Union, United States, Canada and Japan from 2000 to 2009. *Environ. Sci. Technol.* 45: 7232–7240.

Lunestad, B. T. and J. T. Rosnes. 2008. Microbiological quality and safety of farmed fish. In *Improving Farmed Fish Quality and Safety*, ed. L. Øyvind, 199–237. Cambridge: Woodhead Publishing Limited.

Lunestad, B. T., L. Nesse, J. Lassen et al. 2007. *Salmonella* in fish feed; occurrence and implications for fish and human health in Norway. *Aquaculture* 265: 1–8.

Lyle-Fritch, L. P., E. Romero-Beltrán, and F. Páez-Osuna. 2006. A survey on use of the chemical and biological products for shrimp farming in Sinaloa (NW Mexico). *Aquac. Eng.* 35: 135–146.

Maage, A., H. Hove, and K. Julshamn. 2008. Monitoring and surveillance to improve farmed fish safety. In *Improving Farmed Fish Quality and Safety*, ed. L. Øyvind, 547–564. Cambridge: Woodhead Publishing Limited.

MacDonald, D. J. 2005. The HACCP concept and its application in primary production. In *Food Safety Control in the Poultry Industry*, ed. G. C. Mead, 237–254. Cambridge: Woodhead Publishing Limited.

Maule, A. G., A. L. Gannam, and J. W. Davis. 2007. Chemical contaminants in fish feeds used in federal salmonid hatcheries in the USA. *Chemosphere* 67: 1308–1315.

Marty, G. D. 2008. Anisakid larva in the viscera of a farmed Atlantic salmon (*Salmo salar*). *Aquaculture* 279: 209–210.

Midelet-Bourdin, G., S. Copin, G. Leleu, and P. Malle. 2010. Determination of *Listeria monocytogenes* growth potential on new fresh salmon preparations. *Food Control* 21: 1415–1418.

Miettinen, H. and G. Wirtanen. 2006. Ecology of *Listeria* spp. in a fish farm and molecular typing of *Listeria monocytogenes* from fish farming and processing companies. *Int. J. Food Microbiol.* 112: 138–146.

Montory, M. and R. Barra. 2006. Preliminary data on polybrominated diphenyl ethers (PBDEs) in farmed fish tissues (*Salmo salar*) and fish feed in Southern Chile. *Chemosphere* 63: 1252–1260.

Montory, M., E. Habit, P. Fernandez, J. O. Grimalt, and R. Barra. 2012. Polybrominated diphenyl ether levels in wild and farmed Chilean salmon and preliminary flow data for commercial transport. *J. Environ. Sci.* 24: 221–227.

Moss, S. M., D. R. Moss, S. M. Arce, D. V. Lightner, and J. M. Lotz. 2012. The role of selective breeding and biosecurity in the prevention of disease in penaeid shrimp aquaculture. *J. Invertebr. Pathol.* 110: 247–250.

Mozaffarian, D. and E. B. Rimm. 2006. Fish intake, contaminants and human health: Evaluating the risks and the benefits. *J. Am. Med. Assoc.* 296: 1885–1899.

NAMCF. 2008. Response to the questions posed by the Food and Drug Administration and the National Marine Fisheries Service regarding determination of cooking parameters for safe seafood for consumers. *J. Food Prot.* 71: 1287–1308.

Nawaz, M., S. A. Khan, Q. Tran et al. 2012. Isolation and characterization of multi-drug-resistant *Klebsiella* spp. isolated from shrimp imported from Thailand. *Int. J. Food Microbiol.* 155: 179–184.

Nesse, L. L., T. Løvold, B. Bergsjø, K. Nordby, C. Wallace, and G. Holstad. 2005. Persistence of orally administered *Salmonella enterica* serovars Agona and Montevideo in Atlantic salmon. *J. Food Prot.* 68: 1336–1339.

NIFES. 2010. Seafood Data. National Institute of Nutrition and Seafood Research. http://www.nifes.no/sjomatdata/?lang_id=2 (accessed April 17, 2010).

Ohta, S., D. Ishizuka, H. Nishimura et al. 2002. Comparison of ploybrominated diphenyl ethers in fish, vegetables, and meats and levels in human milk of nursing women in Japan. *Chemosphere* 46: 689–696.

O'Toole, S., C. Metcalfe, I. Craine, and M. Gross. 2006. Release of persistent organic contaminants from carcasses of Lake Ontario Chinook salmon (*Oncorhynchus tshawytscha*). *Environ. Pollut.* 140: 102–113.

Petersen, A., J. S. Andersen, T. Kaewmak, T. Somsiri, and A. Dalsgaard. 2002. Impact of integrated fish farming on antimicrobial resistance in a pond environment. *Appl. Environ. Microbiol.* 68: 6036–5042.

Phan, T. T., L. T. L. Khai, N. Ogasawara, N. T. Tam, A. T. Okatani, and M. Akiba. 2005. Contamination of *Salmonella* in retail meats and shrimps in the Mekong Delta, Vietnam. *J. Food Prot.* 68: 1077–1080.

Plessi, M., D. Bertelli, and A. Monzani. 2001. Mercury and selenium content in selected seafood. *J. Food Compost. Anal.* 14: 461–467.

Pruder, G. D. 2004. Biosecurity: Application in aquaculture. *Aquac. Eng.* 32: 3–10.

Radu, S., Noorlis Ahmad, H. L. Foo, and Abdul Reezal. 2003. Prevalence and resistance to antibiotics for *Aeromonas* species from retail fish in Malaysia. *Int. J. Food Microbiol.* 81: 261–266.

Rawn, D. F. K., D. S. Forsyth, J. J. Ryan et al. 2006. PCB, PCDD and PCDF residues in fin and non-fin fish products from the Canadian retail market 2002. *Sci. Total Environ.* 359: 101–110.

Reilly, A. and F. Käferstein. 1997. Food safety hazards and the application of the principles of the hazard analysis and critical control point (HACCP) system for their control in aquaculture production. *Aquac. Res.* 28 (10): 735–752.

Rørvik, L. M. 2000. *Listeria monocytogenes* in the smoked salmon industry. *Int. J. Food Microbiol.* 61: 183–190.

Sapkota, A., A. R. Sapkota, M. Kurcharski et al. 2008. Aquaculture practices and potential human health risks: Current knowledge and future priorities. *Environ. Int.* 34: 1215–1226.

SCAN. 2000. Opinion of the Scientific Committee on Animal Nutrition on the Dioxin Contamination of Feedingstuffs and Their Contribution to the Contamination of Food of Animal Origin. European Commission Health & Consumer Protection Directorate-General, 1-105. http://ec.europa.eu/food/committees/scientific/out55_en.pdf (accessed May 2, 2010).

Schecter, A., O. Päpke, K. C. Tung, D. Staskal, and L. Birnbaum. 2004. Polybrominated diphenyl ethers contamination of United States food. *Environ. Sci. Technol.* 38: 5306–5311.

Shaw, S. D., M. L. Berger, D. Brenner et al. 2008. Polybrominated diphenyl ethers (PBDEs) in farmed and wild salmon marketed in the Northeastern United States. *Chemosphere* 71: 1422–1431.

Sinnott, R. 2002. Fish farming and the feed companies. In *Handbook of Salmon Farming*, ed. S. M. Stead and L. Laird, 105–185. Chichester: Springer Praxis.

Sloth, J. J., K. Julshamn, and A.-K. Lundebye. 2005. Total arsenic and inorganic arsenic content in Norwegian fish feed products. *Aquac. Nutr.* 11: 61–66.

Sohn, H.-Y., C.-S. Kwon, G.-S. Kwon, J.-B. Lee, and E. Kim. 2004. Induction of oxidative stress by endosulfan and protective effect of lipid-soluble antioxidants against endosulfan-induced oxidative damage. *Toxicol. Lett.* 151: 357–365.

Stubhaug, I., L. Frøyland, and B. E. Torstensen. 2005. β-oxidation capacity of red and white muscle and liver in Atlantic salmon (*Salmo salar* L.)—Effects of increasing dietary rapeseed oil and olive oil to replace capelin oil. *Lipids* 40: 39–47.

Sujeewa, A. K. W., A. S. Norrakiah, and M. Laina. 2009. Prevalence of toxic genes of *Vibrio parahaemolyticus* in shrimps (*Penaeus monodon*) and culture environment. *Int. Food Res. J.* 16: 89–95.

Szlinder-Richert, J., I. Barska, J. Mazerski, and Z. Usydus. 2008. Organochlorine pesticides in fish from the southern Baltic Sea: Levels, bioaccumulation features and temporal trends during the 1995-2006 period. *Mar. Pollut. Bull.* 56: 927–940.

Taw, N. T., Y. T. Poh, T. M. Ling, C. Thanabatra, and K. Z. Salleh. 2011. Malaysia shrimp farm redesign successfully combines biosecurity, biofloc technology. *Global Aquacul. Adv.* (March/April): 74–75.

Tendencia, E. A. and L. D. Pena. 2001. Antibiotic resistance of bacteria from shrimp ponds. *Aquaculture* 195: 193–204.

Usydus, Z., J. Szlinder-Richert, L. Polak-Juszczak et al. 2009. Fish products available in Polish market—Assessment of the nutritive value and human exposure to dioxins and other contaminants. *Chemosphere* 74: 1420–1428.

Valdimarsson, G., H. Einarsson, B. Gudbjörnsdottir, and H. Magnusson. 1998. Microbiological quality of Icelandic cooked-peeled shrimp (*Pandalus borealis*). *Int. J. Food Microbiol.* 45: 157–161.

Vaseeharan, B. and P. Ramasamy. 2003. Abundance of potentially pathogenic microorganisms in *Penaeus monodon* larvae rearing systems in India. *Microbiol. Res.* 158: 299–308.

Vogel, B. F., H. H. Huss, B. Ojeniyi, P. Ahrens, and L. Gram. 2001. Elucidation of *Listeria monocytogenes* contamination routes in cold-smoked salmon processing plants detected by DNA-based typing methods. *Appl. Environ. Microbiol.* 67: 2586–2595.

Voorspoels, S., A. Covaci, H. Neelsand, and P. Schepens. 2007. Dietary PBDE intake: A market-basket study in Belgium. *Environ. Int.* 33: 93–97.

WHO. 1999. Food Safety Issues Associated with Products from Aquaculture. Report of a Joint FAO/NACA/WHO Study Group. World Health Organization (WHO Technical Report Series; 883).

WHO. 2002. Use of Antimicrobials Outside Human Medicine and Resultant Antimicrobial Resistance in Humans. Geneva: World Health Organization of the United Nations. http://www.who.int/mediacentre/factsheets/fs268/en/ (accessed April 28, 2010).

WHO. 2006. WHO Guidelines for the Safe Use of Wastewater, Excreta and Greywater, Vol. 3: Wastewater and Excreta Use in Aquaculture. WHO, Geneva, Switzerland. http://whqlibdoc.who.int/publications/2006/9241546832_eng.pdf (accessed January 19, 2010).

Yamashita, Y., Y. Omura, and E. Okazaki. 2005. Total mercury and methyl mercury in commercially important fishes in Japan. *Fish. Sci.* 71: 1029–1035.

Yano, Y., M. Kaneniwa, M. Satomi, H. Oikawa, and S.-S. Chen. 2006. Occurrence and density of *Vibrio parahaemolyticus* in live edible crustaceans from markets in China. *J. Food Prot.* 69: 2742–2746.

Yano, Y., M. Yokoyama, M. Satomi, H. Oikawaand, and S.-S. Chen. 2004. Occurrence of *Vibrio vulnificus* in fish and shellfish available from markets in China. *J. Food Prot.* 67: 1617–1623.

8

Perceptual and Actual Risks and How We Communicate Them

8.1 Introduction

Raw milk. Hand washing. Breastfeeding. What do all these things have in common? Public health officials grappled with the idea of why health care interventions do not work. Behavioral and social science studies are seen as key to the success of public health initiatives (Hayden, 2011). This is because risk perception towards food consumption and food-related hazards are complex. Risks may be associated with both acute and long-term consequences, some of which may have serious effects.

In the previous chapters, we dealt with risk assessments in the technical and natural sciences of food safety risks in the agri-food industries. Each chapter was then followed by prevention and intervention strategies for risk management. In this chapter, we will deal with risk perceptions and communication, which is typically considered within the domain of social sciences and is supported by information from technical risk estimates and consumer perceptions and behaviors.

The concept "risk" means different things to different people (Slovic, 1987). The psychometric approach is used to study people's attitudes to risks and originated from cognitive psychology. It has been utilized to explore public perceptions towards food hazards (Fife-Schaw and Rowe, 1996; Sparks and Shepherd, 1994). The public or laypeople (nonexperts) have a broad conception of risk, which reflects a number of factors such as uncertainty, dread, catastrophic potential, controllability, equity, and risk to future generations (Slovic, 1999). For example, Sparks and Shepherd (1994) found that 87% of the variance in risk perception towards a range of food hazards could be explained by three factors: severity, unknown risks, and number of people exposed. Severity included "concern," "seriousness for future generations," "threatening widespread disastrous consequences," "dread," and "becoming more serious." Unknown risks included "known by people exposed," "known to science," and "accuracy of assessments." The number of people exposed speaks for itself. In the study, dietary and nutritional hazards such as high-fat diet and alcohol consumption were rated low in both the severity and unknown dimension. Since dietary hazards are related to an individual's lifestyle over which he/she has personal control, this may lower the risk scale. Meanwhile, technological hazards such as pesticides and medical residues (over which people felt they have less control) were ranked relatively high in severity and unknown risks (Hansen et al., 2003; Sparks and Shepherd, 1994). In contrast, experts' perceptions of risk are based on the probability of harm or expected mortality, which are synonymous to the makings of risk assessment (risk assessment = likelihood of occurrence × severity of hazard; Slovic, 1999).

At one time, it was widely and straightforwardly assumed that the experts were "right" (Hansen et al., 2003). In fact, trust in experts and in scientific institutions is decreasing, especially after public food scares such as Bovine Spongiform Encephalopathy (BSE), where experts first assured the public that there was no danger to humans and then admitted that this was not the case (Jasanoff, 1997). In designing and implementing risk management strategies, it is important to understand how consumers and experts differ in their risk perceptions. From an economic perspective, the society may benefit through reduction of the substantial economic costs associated with the occurrence of a food safety crisis. Massive loss occurs through product recalls, destruction of implicated food products, loss of incomes, increased hospitalizations and loss of productivity, and increased managerial costs associated with tracking the source of contamination and conducting epidemiological investigations. For example, during the *E. coli* O104:H4 outbreak in Germany in 2011, the German authorities suggested cucumbers from Spain as the source of outbreak. This led to multiple bans across European countries and Russia, causing an estimated loss of 200 million euros ($287 million) a week (Fidler, 2011).

8.2 Dimensions of Risk Perception

Perceived risks are connected to morbidity and mortality along two dimensions: dreaded risk and unknown risk (Figure 8.1). The unknown risk can be described as unfamiliar, uncertain, unknown, uncontrollable, and with delayed consequences (Ueland et al., 2012). The other factors that measure risk perception along the dread axis include being fatal, catastrophic, involuntary, and not easily reduced (Fischhoff et al., 1978; Slovic, 1987).

Foods that are perceived as risky are often foods that are unfamiliar or produced by novel technologies. Foods that are highly processed are also considered to be less desirable and more risky compared to minimally processed food. This is because the possibility to discern what the food product is made of or what it is derived from contributes to a feeling of safety,

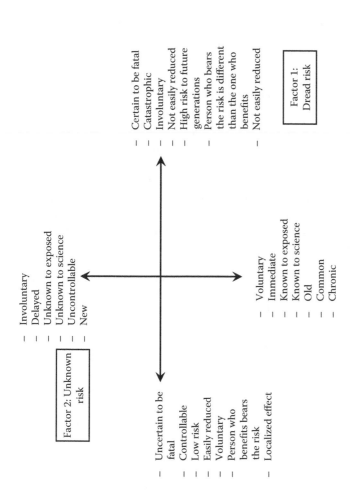

FIGURE 8.1

Dimensions of risk perception: Dread and unknown risk. (Adapted from Fischhoff, B. et al. 1978. *Policy Sci.* 9: 127–152. Slovic, P. 1987. *Science* 236: 280–285; Ueland, Ø. et al. 2012. *Food Chem. Toxicol.* 50: 67–76.)

hence leading to a lower risk perception among consumers (Tijhuis et al., 2012). Consumer concerns about food safety issues tend to focus on chemical substances and novel foods (Bäckström et al., 2003; van Putten et al., 2006), genetically modified food (Frewer et al., 2004; MacFarlane, 2002), food irradiation and pesticide residues (MacFarlane, 2002), and microbiological contaminants such as *Salmonella* spp., *Campylobacter* spp., *Listeria monocytogenes*, and *E. coli* O157:H7 (Wilcock et al., 2004).

Bäckström et al. (2003) also revealed an interesting point in terms of risk perceptions towards novel foods. Novel foods centering on exotic shellfish, raw fish, snails, snakes, and insects evoked a positive response from the focus group panels. Even though the respondents stated that they would not eat the exotic foods, nevertheless, they are perceived as safe since this category of food have already been "safely tested" in another culture.

8.2.1 Lay and Expert Perceptions of Food Safety Risks

In a study by Jensen et al. (2005), the researchers interviewed experts and laypeople to explore how both groups perceive and judge the risk of contracting zoonoses through food. Most lay respondents ranked BSE and variant Creutzfeldt-Jakob disease (vCJD) as the most serious, while experts ranked *Salmonella* and *Campylobacter* as the most serious zoonotic food safety problems. Laypeople ranked *Salmonella* and *Campylobacter* as less serious compared to BSE/vCJD because they can exert a certain level of control over the foodborne pathogens (i.e., by cooking meat and eggs adequately; De Boer et al., 2005). This actually helps to explain why risks thought to be controllable (i.e., preventable by personal action) are likely to evoke a lower risk perception towards certain food hazards. When individuals think "It won't happen to me" and believe that food safety risks are less likely to happen to them than to others, even when the comparison is made with someone from similar demographic characteristics, this is known as *optimistic bias* (Miles and Scaife, 2003; Weinstein, 1980; Weinstein, 1984). Now, we can actually distinguish between two broad categories of potential hazards: (1) those related to technology and its applications (e.g., pesticides, antibiotics, food irradiation, genetically modified food) and (2) those related to lifestyle choices (e.g., individual nutrient consumption, hygiene practices, consumption of rare beef, and sashimi). People are frequently overly optimistic about their own risks from lifestyle hazards.

The experts also took into account the frequency and seriousness of zoonotic infection, for example, how many infections and how serious these infections are. By giving priority to *Salmonella* over BSE/vCJD, an individual can be forced to accept a small risk of death for the sake of reducing the total number of human *Salmonella* infections. Laypeople, however, also consider the consequence of "death" as more serious than the consequence of "illness," whereas experts are willing to make a "proportional" trade-off between these two consequences. This trade-off is considered unacceptable by laypeople, and from a moral viewpoint, an individual also has a right not to have risks placed upon him without his consent. One can say that experts tend to focus on probability while the public pay more attention to consequences.

Public perceptions of risk have often been dismissed on the basis of "irrationality" and tend to be excluded from policy decision making (Frewer et al., 2004). Although experts and decision makers at times are frustrated by the public who worries extensively about small risks, there are some indications that experts may at times actually underestimate risk. Take the probability of an accident such as Chernobyl. It was estimated that such an accident occurs once in 10,000 years while the risk of an accident on the space shuttle was estimated at one in 100,000. Actually, the rate of occurrence turned out to be in a closer range of 1 in 25–50 years (Freudenburg and Pastor, 1992).

However, in order for risk management and risk communication to work in public health and society, the public's risk perception must be taken into consideration. So, how do we resolve the conflict between experts and laypeople's risk perceptions? Experts can respect the lay right by involving the general public in the decision and educate the public to the reality that there is no such thing as absolute safety. Public perceptions of risk are multidimensional and incorporate other factors (e.g., social, cultural, scientific, personal experiences, and political). These factors are rarely included in the risk assessment process and ignored in the risk management decisions (MacFarlane, 2002). In order to provide effective risk management and communication about food safety to the public, it is essential that experts and policymakers are aware of the food safety risk management inadequacies perceived by consumers. This information can then be used to develop improved risk communication and management strategies (Cope et al., 2010). One must remember that regulations can never completely and totally protect the public (Wilcock et al., 2004). Raising awareness among the public may be able to convince the public that it should accept a small risk of death for the sake of the common good in terms of the total number of human infections.

8.2.2 Perceived Risks and Perceived Benefits

In terms of perceived risks and perceived benefits, people will tolerate some degree of risks provided that a benefit is associated with the hazard in question and the benefit accrues to the risk taker (Alhakami and Slovic, 1994). For example, fish and seafood may contain potential sources of environmental contaminants (e.g., polychlorinated biphenyls, dioxins, methyl mercury; Jacobs et al., 2002; Kris-Etherton et al., 2003; Knowles et al., 2003; Hites et al., 2004), and regular consumption of some fish products may result in a negative impact on human health. At the same time, increased consumption of omega-3 fatty acids may be potentially beneficial for health through improved cardiovascular functioning (Kris-Etherton et al., 2002). Perceived risk is important as a motivator of behaviors to avoid, prevent, and manage food safety risks. From a health risk point of view (e.g., cancer), Leventhal et al. (1999) has ably indicated that perceptions of risk are related to motivation to act and to action, and that by increasing the match between perceived risk (beliefs) and actual risk (reality), this will encourage individuals to adopt preventive and treatment behaviors at a level that is appropriate to their actual risk and its source.

8.3 Risk Communication

Risk communication is best described as "The flow of information and risk evaluations back and forth between academic experts, regulatory participants, interest group, and the general public" (Leiss, 1996). Or according to FAO/WHO (1999), "The interactive exchange of information and opinions concerning risk and risk-related factors among risk assessors, risk managers, consumers, and other interested parties."

One of the problems faced in communicating risks originate from the differing languages used to describe risks. Experts tend to use scientific and statistical languages while the public are more intuitive (Table 8.1).

When the public want information about a risk, they prefer a clear message regarding risks and associated uncertainties (Frewer, 2004). For example, in a study by Frewer et al. (2002), people wanted to be provided with information about food risk uncertainty as soon as the uncertainty was identified. People were also more accepting of uncertainty associated with

TABLE 8.1

Experts and Public Assessment of Risk

Expert	Public
An expert will say: "A lifetime intake of aflatoxin at a concentration of twenty parts per billion in food yields an estimated carcinogenic risk to the exposed population (based on 95% confidence level) of one case in a million."	A layperson may say: "Will my children be safe if they eat peanut butter sandwiches every day?"
Scientific	Intuitive
Probabilistic	Yes/no
Acceptable risk	Safety
Changing knowledge	Is it or isn't it?
Comparative risk	Discrete events
Population averages	Personal consequences
A death is a death	It matters how we die

Source: Adapted from Powell, D. and W. Leiss. 1997. *Mad Cows and Mother's Milk: The Perils of Poor Risk Communication*, 26–40, Montreal: McGill-Queen's University Press.

the scientific process of risk management than they were of uncertainty due to lack of action or lack of interest on the part of the risk experts. Meanwhile, experts felt that uncertainties associated with potential risks should not be communicated to the public as they could not understand the implications or may result in overreaction by the public. This resulted in the public perceiving a lack of transparency in the risk analysis process due to the failure of experts to communicate information about risk uncertainty (Cope et al., 2010; Miles and Frewer, 2003). Experts perceived the process of food risk analysis to be adequately transparent, which was not found to be the case for consumers. Improved transparency in the risk analysis implies the need to include information about how food safety activities are initiated and implemented accompanied with risk management strategies and inherent uncertainties (Van Kleef et al., 2007). In other words, effective risk communication should not only include information about the risks associated with food hazards, but also what is being done by risk managers in order to mitigate these risks and what is being done to reduce the uncertainty (Frewer et al., 2002).

In the case of the BSE crisis in the United Kingdom, it has often been suggested that failure to communicate scientific uncertainty associated with the risks resulted in decreased trust in risk management processes and the regulatory institutions that control those processes. In fact, even after advisory experts and policymakers received new information and had changed their views of the risk of BSE to humans, these were never clearly communicated to the public (Jensen, 2004). On the other end, this was in contrast to the case of Swedish acrylamide scare in 2002. Lofstedt (2003) demonstrated that when regulators raised the potential health risks associated with eating fried and baked foods, the public became unnecessarily concerned and confused.

8.3.1 Mass Media

Food safety risk information is usually communicated to the public through media ranging from national mass media to the Internet. There are differences between consumers and experts in their perceptions towards the role of mass media in shaping societal concerns to food risks. The media is viewed by consumers as an important source of information about food safety. As such the media play a role in influencing public perceptions of risk (Reilly and Miller, 1998). The media was viewed by experts as being responsible for public alarm, through "social amplification of risk" (Pidgeon et al., 2003). The social amplification process itself is made possible by the occurrence of a risk-related event or by the potential for a risk-related event (Kasperson et al., 1988). The process occurs when the food safety risk event was selected by a "transmitter" (i.e., mass media or interpersonal network), which either amplifies or attenuates the risk. The transmission is then continued by members of society that may also amplify or attenuate the risk into

TABLE 8.2

Examples of Media Titles

Titles	Media	References
"Killer Cucumber Bug" from Spain Hits Britain	*Sky News,* May 28, 2011	Little (2011)
Preparing for Flu Failure	*Wall Street Journal,* 29 November 2005	Brown (2005)
Oranges Are Not the Safest Fruit—They All Exceed Pesticide Limits	*Independent,* December 18, 2005	Lean (2005)
Your Dinner May Be No Spring Chicken	*The Times,* March 27, 2004	Eliot (2004)
Curry to Dye For	*The Independent,* March 24, 2004	Kirby (2004)
Scottish Farmed Salmon "Is Full of Cancer Toxins"	*The Telegraph,* January 9, 2004	Highfield (2004)
Mad Cow Can Kill You	*Daily Mirror,* March 20, 1996	Cited by Darnton (1996)
Should Ours Be the Only Children in the World to Eat British Beef?	*The Independent,* March 23, 1996	Arthur (1996)

a message, hence resulting in the "ripple effect." Examples of food risk issues which emulate the social amplification of risk are GM food and BSE.

Is it possible to control the flow of information to the public? It is difficult since the media tend to make choices about what is worthy of reporting given the limits on space, time, and audience capacity (Weingart et al., 2000). There is a general perception that the media have a tendency to communicate information on food risks that are misleading and in some cases "sensational." Media publicity at times may also blow the risk out of proportion to maximize impact. Take the following case as an example. The zoonotic capabilities of BSE were widely speculated upon—with estimates of 10,000 cases of the human form of the disease (vCJD) due to consumption of infected beef. However, there were about 150 deaths linked to vCJD. The perceived risk and the fear of this disease appear to far outweigh the actual risk (Smith et al., 2005). Table 8.2 shows examples of titles reported in the media. Numerous food scares had occurred throughout the years, and media play the most important role in disseminating food risk issues.

However, when McCarthy et al. (2008) conducted a snapshot media audit in Ireland, the researchers found that the majority of the newspaper reports highlighting the case of microbiological outbreaks and incidents were nonsensational. The majority of information given in the press releases was represented fairly accurately in the newspaper articles. Prue et al. (2003) also made a similar observation during the U.S. anthrax attacks in the translation of press releases into media articles. A good working relationship between the communicator and media is important in this case and will benefit both stakeholders. McCarthy and Brennan (2009) also suggested that investigations into how this relationship could better operate are essential to the development strategies for food risk issues.

8.3.2 Risk Communication Strategies

Experts noted that the level of education and age were important determinants for the level of understanding of food risk issues and messages. Early intervention via school curricula is the best method to improve public understanding of food risk messages in the long term (De Boer et al., 2005). Hence, the inclusion of food science and food safety (and home economics) into primary and secondary school curricula is desirable (McCarthy and Brennan, 2009).

Remember in Chapter 5 that we discussed the risk factors of dairy farms? Most raw milk drinkers were from the farming communities. When Hegarty et al. (2002) studied the Irish farming families, the authors found out that farmers generally believe their milk to be risk-free on the basis of routine test results. They also believe it to be of better quality than pasteurized milk and also a cheaper option. Meanwhile, palatability, on the other hand, the taste and thickness of raw milk is important for the Hispanic community in California (Headrick et al., 1997). One approach to effective risk communication is to focus on segmenting the population according to their information needs and develop information with high levels of personal relevance to specific groups of respondents (i.e., dairy farm communities; Verbeke et al., 2007). This reiterates the fact that a more informed, public health message needs to be devised to take farming communities' (or specific groups such as the Hispanic population) views into consideration.

8.3.3 Designing Effective Food Safety Messages

The majority of foodborne illnesses are thought to be preventable if food safety principles are understood and practiced throughout the food chain (Jacob et al., 2010). Interventions such as improving food-handling practices and food safety campaigns are necessary to reduce foodborne illnesses (Wong et al., 2004). However, Redmond and Griffith (2003) noticed that despite educational efforts and food safety training, unsafe food-handling practices are still frequently used. Hence, Jacob et al. (2010) suggested that effective food safety messages using new media may effectively modify inappropriate human behaviors in the food safety system (Table 8.3).

A study by Frewer et al. (1997) who reviewed risk communication messages found that statements providing direction, and to which individuals could personally relate, were considered the most persuasive. The messages considered least persuasive included those containing information that were not directly related to the individual, statements with unfamiliar words such as *Campylobacter* and statements which contained unnecessary dates (e.g., "the link between food poisoning and bacteria came in 1888 ..."). It

TABLE 8.3

Factors to Consider When Communicating Food Safety Risks

Factors to Consider When Communicating Food Safety Risks	Comments	References
Understand the target audience	Food safety message should be developed according to audience's needs, concerns, and interests	
Identify appropriate media for distribution	Internet is increasingly being used as a communication tool for food safety and health-related information	Redmond and Griffith (2006)
Challenge complacency	People with an "It will not happen to me" attitude may ignore risk communications, assuming that these messages are targeted at more vulnerable population	Miles et al. (1999)
Enhance personal perception of risk	Emphasize the human rather than the statistical aspects of a story. Identifying individual victims further enhances public perception of personal risk	Joint FAO/WHO Expert Consultation on the Application of Risk Communication to Food Standards and Safety Matters (1998)
Use narratives	Narrative-based messages and messages incorporating fear are more favorably evaluated by farmers than messages that simply inform or that rely on statistics	Morgan et al. (2002)
Associate with audience's lifestyle	Incorporate everyday context into food safety communications. Personal shortcomings such as hunger, lack of money and/or inability to access different foods contribute to different food behaviors	Wilcock et al. (2004)
Reinforce food safety messages	Provide information in written, verbal, or visual formats, but will be most effective if used in combination with each other	Durant (2002)
Use clear language and include graphics	Use clear, nontechnical language appropriate to the target audience and use pictorial materials to clarify messages	
Maintain consistency	Contradictory messages can cause confusion and create distrust in information	
Pretest and evaluate messages	Pretest on target audience on the context in which they will be distributed and revised based on the results	McDermott et al. (2003)

is important that risk communication provides meaningful, relevant, and accurate information in clear and understandable terms targeted to a specific audience (FAO/WHO, 1999). It may also be ineffective to provide vast amounts of information to consumers due to information overload, which leads to confusion and lack of interest among the majority of consumers (Verbeke et al., 2007). Factors such as information overload, irrelevant or useless information, and the need to filter and choose the correct information may result in consumers' reluctance to process information and make decisions (Verbeke, 2005).

8.3.4 Information Sharing and Social Networking

Social media has developed into a number of forms which include text, images, and audio and video sharing through the development of forums, message boards, photo sharing (e.g., Flickr, Instagram), podcasts RSS (really simple syndication), Wikis, social networks (e.g., Facebook, MySpace, Google+), professional networks (e.g., LinkedIn) and microblogging sites (e.g., Twitter, Tumblr) (Wright and Hinson, 2009). Blogs and Twitter are efficient communication tools and are seen as more dialogic, with interactive and faster service for the building of relations (Kent et al., 2003; Schultz et al., 2012). In the time it has taken to read till here, approximately 2000 blogs have been created (Ace Reader, 2012; Sigfry, 2006).

The urgency of providing reliable information to the public is crucial during a crisis. The viral spread of information on social media could be viewed as an effective tool to communicate risks to the public as quickly as possible (Freberg, in press). One good example of the utilization of social media was the recent *E. coli* O104:H4 outbreak of fenugreek seeds in Europe. The rapid dissemination of microbiological findings and technical recommendations during the outbreak should be applauded. This outbreak investigation has demonstrated to the public the transparency and competency of many universities and government institutes which contributed to the identification of the strain (Suerbaum, 2011). Microbiology findings, genome sequencing data, risk assessments, and technical recommendations were released for public access immediately. The rapid dissemination of key data has assisted the public health and food safety laboratory tremendously (Struelens et al., 2011). Rapid risk assessments were conducted and published on the website of ECDC, outbreak updates and microbiological findings were aptly published in *Eurosurveillance*, WHO, and the ECDC-supported Epidemic Intelligence Information System (EPIS) rapid exchange platform. The EPIS links all EU/EEA public health laboratories in the Food and Waterborne Diseases and Zoonoses network (FWD-Net) of the ECDC (Struelens et al., 2011). Teleconferencing was held and aided the sharing of information between health workers. Information sharing through blogs, Twitter, emails and audio podcasts were conducted (Jansen and Kielstein, 2011) and communication via social networking approaches improved collaboration and communication between researchers.

8.4 How to Change Bad Food Handling Behavior

Understanding behavioral science (and changing behavior) is challenging due to multiple internal and external factors. Some examples of internal factors include motivation to change, past experiences, perceived benefits, perceived susceptibility, perceived barriers and attitudes. Extrinsic factors include social pressure to conform, influence from media, environmental factors, and provision of facilities, incentives, or fines. In order to change bad food handling behavior, one needs to understand that provision of knowledge alone is insufficient (Pilling et al., 2008; Soon and Baines, 2012; Seaman and Eves, 2006). In fact, by utilizing social cognitive theory (Figure 8.2), this will enhance the understanding of the complex behavioral interactions of an individual.

8.4.1 Providing Food Safety Knowledge Is Not Sufficient

Studies have shown that increasing knowledge does not necessarily lead to changes in behaviors (Clayton et al., 2002; Ehiri et al., 1997; Rennie, 1994). For example, a study by Byrd-Bredbenner et al. (2007) found that although 97% of the participants rated their own food safety knowledge as at least fair; 60% did not wash their hands with soap and water after touching raw poultry.

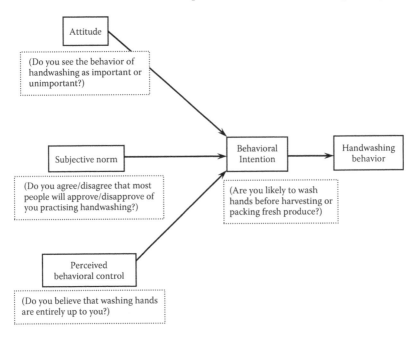

FIGURE 8.2
Theory of Planned Behavior; dotted boxes represent scenarios in a fresh produce farm.

To be effective, food safety messages need to target changing the behavior most likely to result in foodborne illnesses. Most food safety messages and training rely heavily on the provision of information. There is an implied assumption that such training leads to changes in behavior, based on the Knowledge, Attitudes and Practices (KAP) model (Egan et al., 2007). This model has been criticized by Ehiri et al. (1997) and Griffith (2000). Griffith (2000) argued that behavioral change (i.e., the implementation of required hygiene practices) is not easily achieved and that consideration must be given to motivation, constraints, barriers, and facilities as well as to cultural aspects. In order to enhance the effectiveness of risk communication strategies, one must first understand the consumers' and food-handlers' behavior and how this behavior interacts with their beliefs and levels of knowledge. This can be facilitated by use of theory-based models in the development of educational materials (National Cancer Institute, 2005).

The Theory of Planned Behavior (TPB) identifies the influences that predict and change behaviors. Behavioral intention is influenced by: a person's attitudes; beliefs about whether individuals who are important to the person approve or disapprove of the behavior; and perceived control over performing the behavior (National Cancer Institute, 2005; Pilling et al., 2008). Social cognitive models have been used to predict or understand factors underlying individuals' behaviors. TPB and Health Belief Model (HBM) were used to investigate the underlying causes of food safety behaviors within a domestic setting (Clayton et al., 2003). The TPB model was used to determine the impact of food safety training on food handlers in a hospital setting (Seaman and Eves, 2010), food-handling practices among students (Fulham and Mullan, 2011; Mullan and Wong, 2009), and to identify specific beliefs that may be targeted in food safety behaviors (i.e., hand washing, using thermometers, and handling of food contact surfaces) (Pilling et al., 2008).

8.4.2 Evaluation of Hand-Washing Intention among Fresh Produce Farm Workers and Targeting Specific Beliefs

In Figure 8.2 the TPB framework was used to investigate hand-washing intention among fresh produce farm workers. According to the TPB, as a direct determinant of safe food-handling behavior (i.e., hand hygiene practices), the behavioral intention (BI) to adopt hand washing is influenced by three sets of beliefs: attitudes towards the hand-washing practices (defined as an individual's positive or negative evaluation of hand-washing behavior), subjective norms (defined as an individual's perception of the social pressure to perform or not perform the hand-washing behavior), and perceived behavioral control (defined as an individual's perception of the ease or difficulty in performing the hand-washing behavior). Behavioral intention has been identified as the most immediate determinant of behavior (Fishbein and Ajzen, 1975). The advantage of using the TPB framework is that it helps to determine which factor best predicts the intent to perform a behavior

(Shapiro et al., 2011). A better understanding of the beliefs among fresh produce farm workers on their adherence to hand hygiene may help in targeting specific components of the TPB for further improvement.

The components that made up TPB can be targeted in educational interventions to improve behavioral intent. Farm owners and managers should emphasize the positive outcomes of hand washing (e.g., safe produce, less recall, hence more profit for the farm and workers) and potential negative outcomes (causing foodborne illnesses, product recall and lost business, hence affecting staff wages). To improve attitudes, the farm food safety training materials included why food safety is important (e.g., by emphasizing that practicing good hand hygiene reduces the number of people getting sick, through the use of Mason Jones's and Kyle Allgood's case studies during training). In addition, the supervisors and managers should create an environment that cultivates hand washing by putting up posters and reminders (in the workers' native language) and to be role models.

To improve subjective norms, the authors stress that managers, colleagues, health inspectors, and customers would want them to properly perform the behaviors. Motivation from supervisors and management, the support and facilities given to staff are critical to the success of food safety practices. These will contribute to changing attitudes and company culture, and have an impact on behavior and therefore on foodborne outbreaks caused by food workers (Todd et al., 2007).

Perceived control can be improved by supplying adequate resources and reminding employees to perform the behaviors (Pilling et al., 2008). The more accessible the toilets and hand-washing facilities are, the greater the likelihood that they will be used (FDA, 1998); in addition to providing mobile toilets with adequate hand-washing facilities, farm managers can also provide small bottles of hand sanitizers for their staff. However, to move from perceived to actual behavior changes, additional studies on observing fresh produce farm workers' actual hand-washing practices and frequency of washing would be useful. Direct observation is recommended by WHO as the most reliable method for measuring adherence rates to hand hygiene (Boyce and Pittet, 2002) and it is also able to identify the strengths and weaknesses of hand hygiene practices. However, direct observation may result in workers changing their behavior (Hawthorne effect) when they know that they were being observed and can result in falsely elevated compliance rates (Haas and Larson, 2007).

References

Ace Reader. 2012. Read faster and improve concentration. Available at: http://www.acereader.com/ (accessed March 28, 2012).

Alhakami, A. S. and P. Slovic. 1994. A psychological study of the inverse relationship between perceived risk and perceived benefit. *Risk Anal.* 14: 1085–1096.

Arthur, C. 1996. Should ours be the only children in the world to eat British beef? *The Independent*, March 23, 1996. http://www.independent.co.uk/news/should-ours-be-the-only-children-in-the-world-to-eat-british-beef-1343497.html (accessed March 18, 2012).

Bäckström, A., A.-M.Pirttilä-Backman, and H. Tuorila. 2003. Dimensions of novelty: A social representation approach to new foods. *Appetite* 40: 299–307.

Boyce, J. M. and D. Pittet. 2002. Guideline for hand hygiene in health-care settings. *Infect. Control Hosp. Epidemiol.* 23: S23–S40.

Brown, D. 2005. Preparing for flu failure, 11, *Wall St. J.*: November 29.

Byrd-Bredbenner, C., J. Maurer, V. Wheatley, D. Schaffner, C. Bruhn, and L. Blalock. 2007. Food safety self-reported behaviors and cognitions of young adults: Results of a national study. *J. Food Prot.* 70: 1917–1926.

Clayton, D. A., C. J. Griffith, and P. Price. 2003. An investigation of the factors underlying consumers' implementation of specific food safety practices. *Brit. Food J.* 105: 434–453.

Clayton, D. A., C. J. Griffith, P. Price, and A. C. Peters. 2002. Food handler's beliefs and self-reported practices. *Int. J. Environ. Health Res.* 12: 25–39.

Cope, S., L. J. Frewer, J. Houghton, G. Rowe, A. R. H. Fischer, and J. de Jonge. 2010. Consumer perceptions of best practice in food risk communication and management: Implications for risk analysis policy. *Food Policy* 35: 349–357.

Darnton, J. 1996. Britain ties deadly brain disease to cow ailment. *The New York Times*, March 21, 1996. http://www.nytimes.com/1996/03/21/world/britain-ties-deadly-brain-disease-to-cow-ailment.html?pagewanted=all&src=pm (accessed March 18, 2012).

De Boer, M., M. McCarthy, M. Brennan, A. L. Kelly, and C. Ritson. 2005. Public understanding of food risk issues and food risk messages on the island of Ireland: The views of food safety experts. *J. Food Saf.* 25: 241–265.

Durant, D. 2002. Take a look at real magic: Disney and food safety. *The Food Safety Educator*, 7(2).http://www.fsis.usda.gov/Frame/FrameRedirect.asp?main = http://www.fsis.usda.gov/OA/educator/educator7-2.htm (accessed March 18, 2012).

Egan, M. G., M. M. Raats, S. M. Grubb et al. 2007. A review of food safety and food hygiene training studies in the commercial sector. *Food Control* 18: 1180–1190.

Ehiri, J. E., G. P. Morris, and J. McEwen. 1997. Evaluation of a food hygiene training course in Scotland. *Food Control* 8: 137–147.

Eliot, V. 2004. Your dinner may be no spring chicken, 3, *The Times*, March 27.

FAO/WHO. 1999. The application of risk communication to food standards and safety matters. *Report of a Joint FAO/WHO Expert Consultation Rome*, February 2–6, 1998. http://www.fao.org/docrep/005/x1271e/x1271e00.htm (accessed March 18, 2012).

FDA. 1998. Guide to minimize microbial food safety hazards for fresh fruits and vegetables. Food and Drug Administration, U.S. Department of Agriculture, Centers for Disease Control and Prevention. http://www.fda.gov/downloads/Food/GuidanceComplianceRegulatoryInformation/GuidanceDocuments/ProduceandPlanProducts/UCM169112.pdf (accessed January 9, 2011).

Fidler, D. P. 2011. International law and the *E. coli* outbreaks in Europe. *ASIL Insights* 15: 14.

Fife-Schaw, C. and G. Rowe. 1996. Public perceptions of everyday food hazards: A psychometric study. *Risk Anal.* 16: 487–500.

Fischhoff, B., P. Slovic, and S. Lichtenstein. 1978. How safe is safe enough? A psychometric study of attitudes towards technological risks and benefits. *Policy Sci.* 9: 127–152.

Fishbein, M. and I. Ajzen. 1975. *Belief, Attitude, Intention and Behavior: An Introduction to Theory and Research.* Reading: Addison-Wesley.

Freberg, K. In press. Intention to comply with crisis messages communicated via social media. *Public Relat. Rev.*

Freudenburg, W. R. and S. K. Pastor. 1992. NIMBYs and LULUs: Stalking the syndromes. *J. Soc. Issues* 48: 39–61.

Frewer, L. 2004. The public and effective risk communication. *Toxicol. Lett.* 149: 391–397.

Frewer, L. J., C. Howard, D. Hedderley, and R. Shepherd. 1997. The elaboration likelihood model and communication about food risks. *Risk Anal.* 17: 759–770.

Frewer, L., J. Lassen, B. Kettlilz, J. Scholderer, V. Beekman, and K. G. Berdal. 2004. Societal aspects of genetically modified foods. *Food Chem. Toxicol.* 42: 1181–1193.

Frewer, L. J., S. Miles, M. Brennan, S. Kuznesof, M. Nessand, and C. Ritson. 2002. Public preferences for informed choice under conditions of risk uncertainty. *Public Underst. Sci.* 11: 363–372.

Fulham, E. and B. Mullan. 2011. Hygienic food handling behaviours: Attempting to bridge the intention-behavior gap using aspects from temporal self-regulation theory. *J. Food Prot.* 74: 925–932.

Griffith, C. 2000. Food safety in catering establishments. In *Safe Handling of Foods*, ed. J. M. Farber and E. C. D. Todd, 235–256. New York: Marcel Dekker.

Haas, J. P. and E. L. Larson. 2007. Measurement of compliance with hand hygiene. *J. Hosp. Infect.* 66: 6–14.

Hansen, J., L. Holm, L. Frewer, P. Robinson, and P. Sandøe. 2003. Beyond the knowledge deficit: Recent research into lay and expert attitudes to food risks. *Appetite* 41: 111–121.

Hayden, E. C. 2011. Aid organizations tap into social-science expertise. *Nature* 479: 163.

Headrick, M. L., B. Timbo, K. C. Klontz, and S. B. Werner. 1997. Profile of raw milk consumers in California. *Public Health Rep.* 112: 418–422.

Hegarty, H., M. B. O'Sullivan, J. Buckley, and C. Foley-Nolan. 2002. Continued raw milk consumption on farms: Why? *Commun. Dis. Public Health* 5: 151–156.

Highfield, R. 2004. Scottish farmed salmon "is full of cancer toxins." *The Telegraph*, January 9, 2004.

Hites, R. A., J. A. Foran, D. O. Carpenter, M. C. Hamilton, B. A. Knuthand, and S. J. Schwager. 2004. Global assessment of organic contaminants in farmed salmon. *Science* 303: 226–229.

Jacob, C., L. Mathiasen, and D. Powell. 2010. Designing effective messages for microbial food safety hazards. *Food Control* 21: 1–6.

Jacobs, M., J. Ferrario, and C. Byrne. 2002. Investigation of polychlorinated dibenzo-*p*-dioxins, dibenzo-*p*-furans and selected coplanar biphenyls in Scottish farmed Atlantic salmon (*Salmo salar*). *Chemosphere* 47: 183–191.

Jansen, A. and J. T. Kielstein. 2011. The new face of enterohaemorrhagic *Escherichia coli* infections. *Eurosurveill.* 16: pii = 19898. http://www.eurosurveillance.org/images/dynamic/EE/V16N25/art19898.pdf (accessed November 16, 2011).

Jasanoff, S. 1997. Civilization and madness: The great BSE scare of 1996. *Public Underst. Sci.* 6: 221–232.

Jensen, K. K. 2004. BSE in the UK: Why the risk communication strategy failed. *J. Agric. Environ. Ethics* 17: 405–423.

Jensen, K. K., J. Lassen, P. Robinson, and P. Sandøe. 2005. Lay and expert perceptions of zoonotic risks: Understanding conflicting perspectives in the light of moral theory. *Int. J. Food Microbiol.* 99: 245–255.

Joint FAO/WHO Expert Consultation on the Application of Risk Communication to Food Standards and Safety Matters. 1998. The application of risk communication to food standards and safety matters, February 2–6, Rome, Italy. ftp://ftp. fao.org/docrep/fao/005/x1271e/x1271e00.pdf (accessed March 18, 2012).

Kasperson, R. E., O. Renn, P. Slovic et al. 1988. The social amplification of risk: A conceptual framework. *Risk Anal.* 8: 177–187.

Kent, M. L., M. Taylor, and W. J. White. 2003. The relationship between web site design and organizational responsiveness to stakeholders. *Public Relat. Rev.* 29: 63–77.

Kirby, T. 2004. Curry to dye for. *The Independent*, March 24, 2004. http://www.independent.co.uk/life-style/health-and-families/health-news/curry-to-dye-for-567453.html (accessed March 18, 2012).

Knowles, T. G., D. Farrington, and S. C. Kestin. 2003. Mercury in UK imported fish and shellfish and UK-farmed fish and their products. *Food Addit. Contam.* 20: 813–818.

Kris-Etherton, P. M., W. S. Harrisand, and L. J. Appel. 2003. Fish consumption, fish oil, omega-3 fatty acids and cardiovascular disease. *Arterioscl. Throm. Vas.* 23: e20–e30.

Lean, G. 2005. Oranges are not the safest fruit—they all exceed pesticide limits. *The Independent,* 18 December. http://www.independent.co.uk/environment/oranges-are-not-the-safest-fruit—they-all-exceed-pesticide-limits-519954.html (accessed March 18, 2012).

Leiss, W. 1996. Three phases in the evolution of risk communication practice. *Ann. Am. Acad. Pol. Soc. Sci.* 545: 85–94.

Leventhal, H., K. Kelly, and E. A. Leventhal. 1999. Population risk, actual risk, perceived risk and cancer control: A discussion. *J. Natl. Cancer Inst. Monogr.* 25: 81–85.

Little, D. 2011. Killer cucumber' bug from Spain hits Britain, *Sky News,* May 28. http://news.sky.com/home/uk-news/article/16001083 (accessed March 18, 2012).

Lofstedt, R. E. 2003. Science communication and the Swedish acrylamide alarm. *J. Health Commun.* 8: 407–432.

MacFarlane, R. 2002. Integrating the consumer interest in food safety: The role of science and other factors. *Food Policy* 27: 65–80.

McCarthy, M. and M. Brennan. 2009. Food risk communication: Some of the problems and issues faced by communicators on the Island of Ireland (IOI). *Food Policy* 34: 549–556.

McCarthy, M., M. Brennan, M. De Boer, and C. Ritson. 2008. Media risk communication—what was said by whom and how was it interpreted. *J. Risk Res.* 11: 375–394.

McDermott, M. H., Chess, C., Perez-Lugo, M., Pflugh, K. K., Bochenek, E., and Burger, J. 2003. Communicating a complex message to the population most at risk: An outreach strategy for fish consumption advisories. *Appl. Environ. Educ. Commun.* 2: 23–37.

Miles, S., D. S. Braxton, and L. J. Frewer. 1999. Public perceptions about microbiological hazards in food. *Brit. Food J.* 101: 744–762.

Miles, S. and L. J. Frewer. 2003. Public perception of scientific uncertainty in relation to food hazards. *J. Risk Res.* 6: 267–283.

Miles, S. and V. Scaife. 2003. Optimistic bias and food. *Nutr. Res. Rev.* 16: 3–19.

Morgan, S. E., H. P. Cole, T. Struttmann, and L. Piercy. 2002. Stories or statistics? Farmers' attitudes toward messages in an agricultural safety campaign. *J. Agric. Saf. Health* 8: 225–239.

Mullan, B. A. and C. L. Wong. 2009. Hygienic food handling behaviours: An application of the theory of planned behaviour. *Appetitie* 52: 757–761.

National Cancer Institute. 2005. *Theory at a Glance. A Guide for Health Promotion Practice.* U.S. Department of Health and Human Services. http://www.cancer.gov/PDF/481f5d53-63df-41bc-bfaf-5aa48ee1da4d/TAAG3.pdf (accessed May 13, 2010).

Pidgeon, N., R. E. Kasperson, and P. Slovic. 2003. *The Social Amplification of Risk.* Cambridge: Cambridge University Press.

Pilling, V. K., L. A. Brannon, C. W. Shanklin, A. D. Howells, and K. R. Roberts. 2008. Identifying specific beliefs to target to improve restaurant employees' intentions for performing three important food safety behaviours. *J. Am. Diet. Assoc.* 108: 991–997.

Powell, D. and W. Leiss. 1997. A diagnostic for risk communication failures. In *Mad Cows and Mother's Milk: The Perils of Poor Risk Communication*, 26–40. Montreal: McGill-Queen's University Press.

Prue, C. E., C. Lackey, L. Swenarskiand, and J. M. Gantt. 2003. Communication monitoring: Shaping CDC's emergency risk communication efforts. *J. Health Commun. Int. Pers.* 8: 35–49.

Redmond, E. C. and C. J. Griffith. 2003. Consumer food handling in the home: A review of food safety studies. *J. Food Prot.* 66: 130–161.

Redmond, E. C. and C. J. Griffith. 2006. Assessment of consumer food safety education provided by local authorities in the UK. *Brit. Food J.* 108: 732–752.

Reilly, J. and D. Miller. 1998. Scaremonger or scapegoat? The role of the media in the emergence of food as a social issue. In *Food, Health and Identity*, ed. P. Caplan, 234–251. New York: Routledge.

Rennie, D. M. 1994. Evaluation of food hygiene education. *Brit. Food J.* 96: 20–25.

Schultz, F., S. Utz, and A. Göritz. 2011. Is the medium the message? Perceptions of and reactions to crisis communication via twitter, blogs and traditional media. *Public Relat. Rev.* 37: 20–27.

Seaman, P. and A. Eves. 2006. The management of food safety—the role of food hygiene training in the UK service sector. *Hospitality Manage.* 25: 278–296.

Seaman, P. and E. Eves. 2010. Efficacy of the theory of planned behaviour model in predicting safe food handling practices. *Food Control* 21: 983–987.

Shapiro, M. A., N. Porticella, L. C. Jiang, and R. Gravani. 2011. Predicting intentions to adopt safe home food handling practices: Applying the theory of planned behavior. *Appetite* 56: 96–103.

Sigfry, D. 2006. State of the blogosphere: On blogosphere growth. http://www.sifry.com/alerts/archives/000419.html (accessed March 28, 2012).

Slovic, P. 1987. Perception of risk. *Science* 236: 280–285.

Slovic, P. 1999. Trust, emotion, sex, politics, and science: Surveying the risk-assessment battlefield. *Risk Anal.* 19: 689–701.

Smith, K. R., P. Clayton, B. Stuart, K. Myers, and P. M. Seng. 2005. The vital role of science in global policy decision-making: An analysis of past, current, and forecasted trends and issues in global red meat trade and policy. *Meat Sci.* 71: 150–157.

Soon, J. M. and R. N. Baines. 2012. Food safety training and evaluation of handwashing intention among fresh produce farm workers. *Food Control* 23: 437–448.

Sparks, P. and R. Shepherd. 1994. Public perceptions of the potential hazards associated with food production and food consumption: An empirical study. *Risk Anal.* 14: 799–806.

Struelens, M. J., D. Palm, and J. Takkinen. 2011. Enteroaggregative, Shiga toxin-producing *Escherichia coli* O104:H4 outbreak: new microbiological findings boost coordinated investigations by European public health laboratories. Eurosurveill. 16: pii = 19890. http://www.eurosurveillance.org/images/dynamic/EE/V16N24/art19890.pdf (accessed November 16, 2011).

Suerbaum, S., 2011. No tech gaps in *E. coli* outbreak. *Nature* 476: 33.

Tijhuis, M. J., M. V. Pohjola, H. Gunnlaugsdóttir et al. 2012. Looking beyond borders: Integrating best practices in benefit-risk analysis into the field of food and nutrition. *Food Chem. Toxicol.* 50: 77–93.

Todd, E. C. D., J. D. Greig, C. A. Bartleson, and B. S. Michaels. 2007. Outbreaks where food workers have been implicated in the spread of foodborne disease. Part 3. Factors contributing to outbreaks and description of outbreak categories. *J. Food Prot.* 70: 2199–2217.

Ueland, Ø., H. Gunnlaugsdottir, F. Holm et al. 2012. State of the art in benefit-risk analysis: Consumer perception. *Food Chem. Toxicol.* 50: 67–76.

Van Kleef, E., J. R. Houghton, A. Krystallis et al. 2007. Consumer evaluations of food risk management quality in Europe. *Risk Anal.* 27: 1565–1580.

Van Putten, M. C., L. J. Frewer, L. J. W. J. Gilissen, B. Gremmen, A. A. C. M. Peijnenburg, and H. J. Wichers. 2006. Novel foods and food allergies: A review of the issues. *Trends Food Sci. Technol.* 17: 289–299.

Verbeke, W. 2005. Agriculture and the food industry in the information age. *Eur. Rev. Agric. Econ.* 32: 347–368.

Verbeke, W., L. J. Frewer, J. Scholderer, and H. F. De Brabander. 2007. Why consumers behave as they do with respect to food safety and risk information. *Anal. Chim. Acta* 586: 2–7.

Weingart, P., A. Engels, and P. Pansegrau. 2000. Risks of communication: Discourses on climate change in science, politics, and the mass media. *Public Underst. Sci.* 9: 261–283.

Weinstein, N. D. 1980. Unrealistic optimism about future life events. *J. Pers. Soc. Psychol.* 39: 806–820.

Weinstein, N. D. 1984. Why it won't happen to me: Perceptions of risk factors and susceptibility. *Health Psychol.* 3: 431–457.

Wilcock, A., M. Pun, J. Khanona, and M. Aung. 2004. Consumer attitudes, knowledge and behaviour: A review of food safety issues. *Trends Food Sci. Technol.* 15: 56–66.

Wong, S., R. Marcus, M. Hawkins et al. 2004. Physicians as food-safety educators: A practices and perceptions survey. *Clin. Infect. Dis.* 38: S212–S218.

Wright, D. K. and M. D. Hinson. 2009. An updated look at the impact of social media on public relations practice. *Public Relat. J.* 3: 1–27.

Index